The UMAP Journal

Modules

Tools for Teaching 1995

published by

The Consortium for Mathematics
and Its Applications, Inc.
Suite 210
57 Bedford St.
Lexington, MA 02173

edited by

Paul J. Campbell
Campus Box 194
Beloit College
700 College St.
Beloit, WI 53511–5595
campbell@beloit.edu

ISBN 0–912843–41–1
Typeset and printed in the U.S.A.

Table of Contents

Erratum

The back cover of this issue incorrectly cites Robert E. D. Woolsey as the author of Unit 746, "How to Win at Nim." The module was written by Daniel E. Loeb.

COMAP

Introduction

The instructional Modules in this volume were developed by the Undergraduate Mathematics and Its Applications (UMAP) Project. Project UMAP develops and disseminates instructional modules and expository monographs in mathematical modeling and applications of the mathematical sciences, for undergraduate students and their instructors.

UMAP Modules are self-contained (except for stated prerequisites) lesson-length instructional units. From them, undergraduate students learn professional applications of the mathematical sciences. UMAP Modules feature different levels of mathematics, as well as various fields of application, including biostatistics, economics, government, earth science, computer science, and psychology. The Modules are written and reviewed by instructors in colleges and high schools throughout the United States and abroad, as well as by professionals in applied fields.

UMAP was originally funded by grants from the National Science Foundation to the Education Development Center, Inc. (1976–1983) and to the Consortium for Mathematics and Its Applications (COMAP) (1983–1985). In order to capture the momentum and success beyond the period of federal funding, we established COMAP as a nonprofit educational organization. COMAP is committed to the improvement of mathematics education, to the continuing development and dissemination of instructional materials, and to fostering and enlarging the network of people involved in the development and use of materials. In addition to involvement at the college level through UMAP, COMAP is engaged in science and mathematics education in elementary and secondary schools, teacher training, continuing education, and industrial and government training programs.

In addition to this annual collection of UMAP Modules, other college-level materials distributed by COMAP include individual Modules (more than 500), *The UMAP Journal*, and UMAP expository monographs. Thousands of instructors and students have shared their reactions to the use of these instructional materials in the classroom, and comments and suggestions for changes are incorporated as part of the development and improvement of materials.

This collection of Modules represents the spirit and ability of scores of volunteer authors, reviewers, and field-testers (both instructors and students). The substance and momentum of the UMAP Project comes from the thousands of individuals involved in the development and use of UMA instructional materials. COMAP is very interested in receiving information on the use of Modules in various settings. We invite you to call or write for a catalog of available materials, and to contact us with your ideas and reactions.

Sol Garfunkel, COMAP Director
Paul J. Campbell, Editor

Recruitment, editing, and selection UMAP Modules is done by the board of editors of *The UMAP Journal*, who are appointed by the editor-in-chief in consultation with the presidents of the cooperating organizations

Mathematical Association of America (MAA),
Society for Industrial and Applied Mathematics (SIAM),
National Council of Teachers of Mathematics (NCTM),
American Mathematical Association of Two-Year Colleges (AMATYC),
Institute for Operations Research and the Management Sciences (INFORMS), and
American Statistical Association (ASA).

In 1995 the editor and associate editors were:

Manuscripts are read double-blind by two or more referees, including an associate editor. Guidelines are published in *The UMAP Journal*.

UMAP

Modules in Undergraduate Mathematics and Its Applications

Published in cooperation with the Society for Industrial and Applied Mathematics, the Mathematical Association of America, the National Council of Teachers of Mathematics, the American Mathematical Association of Two-Year Colleges, The Institute of Management Sciences, and the American Statistical Association.

Module 737

Geometric Programming

Robert E.D. Woolsey

Applications of Optimization and Geometry to Engineering Design

COMAP, Inc., Suite 210, 57 Bedford Street, Lexington, MA 02173 (617) 862–7878

INTERMODULAR DESCRIPTION SHEET: UMAP Unit 737

TITLE: Geometric Programming

AUTHORS: Robert E.D. Woolsey
Mathematical and Computer Sciences Dept.
Colorado School of Mines
Golden, CO 80401

MATHEMATICAL FIELD: Optimization, geometry

APPLICATION FIELD: Engineering design

TARGET AUDIENCE: Students in a course in calculus.

ABSTRACT: What if someone told you that there was a method of optimization that:

- told you how much an engineering design would cost *before* you knew what it looked like.
- After you found out how much it cost, it would *then* tell you what the values of the design variables were.
- It would enable you to calculate the effects of price changes in materials, inflationary effects, or changes in requirements *without solving the problem again.*
- Once you had solved an engineering design with the method, *you never had to do it again*!

And what if they told you that you *didn't need calculus to do the optimization*. Now, we all remember how much *fun* it was to memorize all of those differentials and integrals. With this method, there are only *four rules* to remember, *one sentence long*. If you like the sound of this, *read on*!

PREREQUISITES: Differential calculus, including Lagrange multipliers (so as to appreciate by comparison the simplicity of the approach presented here).

COMAP, Inc., Suite 210, 57 Bedford Street, Lexington, MA 02173
(800) 77–COMAP = (800) 772–6627 (617) 862–7878

1. Introduction

What if someone told you that there was a method of optimization that:

- told you how much an engineering design would cost *before* you knew what it looked like.
- After you found out how much it cost, it would *then* tell you what the values of the design variables were.
- It would enable you to calculate the effects of price changes in materials, inflationary effects, or changes in requirements *without solving the problem again.*
- Once you had solved an engineering design with the method, *you never had to do it again!*

And what if they told you that you *didn't need calculus to do the optimization.* Now, we all remember how much *fun* it was to memorize all of those differentials and integrals. With this method, there are only *four rules* to remember, *one sentence long.* If you like the sound of this, *read on!*

2. A Sample Problem

Dryhole Oil and Refining Company has a cylindrical oil-storage tank that holds 3141.59 cubic yards of oil. Since the company is massively underinsured, it naturally catches fire and burns to the ground in 2.5×10^{-4} seconds. The insurance company, in a burst of generosity, pays off the $1,000 policy instantaneously. The design engineers solicit bids for reconstruction. The low bidder, a subsidiary of the oil company, bids a price of $1 per square yard of surface steel. Before the design engineer is allowed to start construction, he must convince the auditors that the cost will be less than the insurance payoff of $1,000.

Now, the above massively oversimplified problem is the basis for all that is to follow. Let us now approach the problem with classical mathematics. Letting R be the radius of the tank and H its height (both measured in yards), the mathematical formulation might be as follows:

$$\text{MINIMIZE } \pi R^2 \text{ (surface of the top) } + 2\pi RH \text{ (surface of the side),}$$

subject to the constraint that the designed tank must hold 3141.59 cubic yards of oil, or

$$\pi R^2 H \geq 3141.59.$$

We now reformulate the problem using Lagrange multipliers, so that the objective function becomes

$$L = \pi R^2 + 2\pi RH - \lambda\left(\pi R^2 H - 3141.59\right).$$

Differentiating partially with respect to R, H, and λ, we have:

$$\begin{aligned}
\frac{\partial L}{\partial R} &= 2\pi RH + 2\pi H - 2\pi RH\lambda &&= 0,\\
\frac{\partial L}{\partial H} &= 2\pi R - \pi R^2\lambda &&= 0,\\
\frac{\partial L}{\partial \lambda} &= \pi R^2 H - 1000\pi &&= 0.
\end{aligned}$$

These equations can be rewritten as

$$\begin{aligned}
R + H - RH\lambda &= 0,\\
\lambda &= \frac{2}{R},\\
R^2 H &= 1000.
\end{aligned}$$

From the first two, we conclude

$$R + H - 2H = 0, \qquad \text{or} \qquad R = H,$$

while from the third we get

$$R = H = 10.$$

Substituting into the objective function, we get the total cost to be

$$\$^* = \pi(10)^2 + 2\pi(10)(10) = 300\pi = \$942.48.$$

Therefore, the answer is "YES," we can afford to rebuild the tank.

Now let's look at all of the pain and agony that we had to go through to solve the above really simple nonlinear engineering design problem. We had to construct a set of three *nonlinear* equations and solve them. Clearly, if we had a somewhat more complex problem—with, say, 30 variables and 30 equations—we might have a *little* more difficulty. Doing multidimensional search over 30 variables on a computer is no joke, even with well-behaved functions.

Now, what if I told you that there was a way to solve such problems *without* calculus, which required you to to solve, at worst, a set of *linear* equations? And further, what if I told you that the method will give you the *cost* of the design before you have *any* idea of what the wretched thing looks like? And *then* I will tell you that if you ever design anything with this method, using a given model, that *you never have to solve that model again*! Finally, I will tell you that for the great majority of engineering design problems, using this method, the problem, once solved, is independent of changes in any of the cost coefficients, as might come from inflationary effects, wrong data, taxes, or increases in labor or material cost. You would probably find that hard to believe, but read on for the verification of all that I have promised.

This methodology, called *geometric programming* (GP), has the unnerving property of finding the optimal cost *before* you know the optimal values of the variables. Note that in our example, we didn't even yet find the optimal values for the variables. Put another way:

With GP, you can find the cost of a design before designing it.

3. Woolsey's Four Rules for Geometric Programming

To formulate our problem as a GP problem, we must first cast all constraints into the form

$$f(x) \leq 1.$$

The volume constraint now appears as

$$\pi R^2 H \geq 1000\pi,$$

which may be quickly rewritten as

$$1000 R^{-2} H^{-1} \leq 1.$$

The mathematical formulation for the problem as a GP problem is then

$$\text{MINIMIZE} \quad \pi R^2 + 2\pi R H \qquad \textbf{(1)}$$

$$\text{subject to} \quad 1000 R^{-2} H^{-1} \leq 1. \qquad \textbf{(2)}$$

The form of the optimal solution may be written by using the four rules that follow.

Rule 1: The Optimal Form of Any Polynomial G.P. Problem Is:

$$\begin{aligned}
\$^* = {} & \left(\frac{\text{first coeff. of obj. function}}{d_1}\right)^{d_1} \cdots \left(\frac{\text{last coeff. of obj. function}}{d_{\text{last}}}\right)^{d_{\text{last}}} \\
& \times \left\{ \left(\frac{\text{coeff. of first term of first constraint}}{d_{\text{last}+1}}\right)^{d_{\text{last}+1}} \cdots \right. \\
& \times \left(\frac{\text{coeff. of last term of first constraint}}{d_{\text{last}+m}}\right)^{d_{\text{last}+m}} \\
& \left. \times \left(\sum d_i \text{ for first constraint}\right)^{(\sum d_i \text{ for first constraint})} \right\} \\
& \times \cdots
\end{aligned}$$

where the expression in curly brackets is to be repeated for each constraint, and where it is assumed that there are m *terms* in the constraint. A term in GP is anything separated by a $+$, $-$, or an inequality that has a constant part *and* a variable part.

Rule 2: The Exponent Matrix Is Constructed As Follows:

Rule 2A: The sum of contributions to cost in the objective function is:

$$d_1 + d_2 + \cdots + d_{\text{last}} = 1.$$

Rule 2B: For each primal variable, the equations in the exponent matrix are:

$$(\text{power of variable } I \text{ in term } 1)d_1 + (\text{power of variable } I \text{ in term } 2)d_2 + (\text{power of variable } I \text{ in last term})d_{\text{last}+m} = 0.$$

Rule 3: At Optimality, for the Objective Function:

$$TEC^* = \left(\frac{\text{first term of obj. function}}{d_1^*}\right)^{d_1} = \left(\frac{\text{second term of obj. function}}{d_2^*}\right)^{d_1} \left(\frac{\text{last term of obj. function}}{d_{\text{last}}^*}\right)^{d_1}.$$

Rule 4: At Optimality, for Each Constraint:

$$d_I = (I\text{th term of constraint}) \cdot \left(\sum d_I \text{ for that constraint}\right).$$

From Rule 1 and **(1)** and **(2)**, we have at once that the total cost is

$$\$^* = \left(\frac{\pi}{d_1}\right)^{d_1} \left(\frac{2\pi}{d_2}\right)^{d_2} \left(\frac{1000}{d_3}\right)^{d_3} d_3. \quad \textbf{(3)}$$

A good way to remember Rule 1 is to think in terms of rap music. If you were to snap your fingers in four-four time, you might formulate the rule as:

The form of the optimal solution is:

"Coefficient of the first term over delta one to the delta times coefficient of the second term over delta two to the delta two times coefficient of the third term over delta three to the delta three
and
if you got a constraint you gotta multiply by the
sum of the deltas to the sum of the deltas. (YEAH!!)"

(Grateful acknowledgment is made to M.C. Hammer.)

If we knew what the deltas were, we would know the minimum cost *right now*! We must now use Rules 2A and 2B to define the matrix of exponents as follows.

Geometric programming first requires the sums of the deltas, or deltas in the objective function, to be equal to 1. If we think of the deltas as *percentage contributions to cost at optimality*, it should be reasonable that the sum of contributions to cost should equal to 100%. So, for this problem we have:

$$d_1 + d_2 = 1. \tag{4}$$

Rule 2B generates an equation for each variable. We have two variables; we get for the variable R

$$(2) \cdot d_1 + (1) \cdot d_2 + (-2) \cdot d_3 = 0. \tag{5}$$

You could snap your fingers, get a good rhythm going, and rap out:

> "Power of the first variable in the first term times delta one, plus power of the first variable in the second terms times delta two, plus power of the first variable in the third term times delta three gotta be zero!"

Doing the same thing for the variable H, we would get

$$(0) \cdot d_1 + (1) \cdot d_2 + (-1) \cdot d_3 = 0. \tag{6}$$

Once again, snap your fingers, get a good rhythm going, and rap out:

> "Power of the second variable in the first term times delta one, plus power of the second variable in the second term times delta two, plus power of the second variable in the third term times delta three gotta be zero!"

Assembling all of the equations **(4)–(6)**, we get

$$\begin{array}{rcrcrcl} d_1 & + & d_2 & + & 0d_3 & = & 1, \\ 2d_1 & + & d_2 & - & 2d_3 & = & 0. \\ 0d_1 & + & d_2 & - & d_3 & = & 0. \end{array}$$

Note that these three equations are *linear*. This means that their solution is obtained by simply inverting a matrix, something that computers do like *gangbusters*. We quickly determine that the right answers are

$$d_1 = \frac{1}{3}, \qquad d_2 = \frac{2}{3}, \qquad d_3 = \frac{2}{3}.$$

By substitution into the objective function **(3)**, we get the cost expressed as

$$\$^* = \left(\frac{\pi}{1/3}\right)^{1/3} \left(\frac{2\pi}{2/3}\right)^{2/3} (1000)^{2/3} = 300\pi = \$942.48.$$

Notice that:

- Optimal cost was found *first, without calculus.*
- Optimal cost was found *before* finding optimal dimensions.
- A set of *linear* equations was solved, rather than *non*linear ones.

The reader should pause at this point to consider the extension of this approach to *really grim* engineering design problems. We have found that we can afford to rebuild the tank. *Now* we will find the optimal dimensions, using Rule 3. Let us recall it from above:

$$\$^* = \left(\frac{\text{first term of obj. function}}{d_1^*}\right)^{d_1} = \left(\frac{\text{second term of obj. function}}{d_2^*}\right)^{d_1} = \dots$$
$$= \left(\frac{\text{last term of obj. function}}{d_{\text{last}}^*}\right)^{d_1}.$$

Using this and **(1)**, or the cost expressed as $(\pi R^2 + 2\pi RH)$, we have the optimal cost as

$$\$^* = 300\pi = \frac{\pi R^2}{1/3} = \frac{2\pi RH}{2/3},$$

which may be rewritten as

$$\$^* = 300\pi = 3\pi R^2 = 3\pi RH.$$

Note that by looking at the contribution of the first and second terms, we get $R = 10$. And by looking at the combination of the second and third terms, we get $R = H$. We now recall Pythagoras's beloved rule that "things equal to the same thing are equal to each other " and get, finally, that $H = 10$. The solution is now complete.

But wait! GP has delivered something extra. Let's look again at the optimal deltas: $d_1 = 1/3$ and $d_2 = 2/3$. The determination of the deltas *has nothing to do with the actual costs of the problem*! The deltas express only the percentage contribution of their terms to the total cost, *regardless of the unit cost term*. In short, the deltas tell us that the *top* of the optimally designed tank will contribute *one-third* of the total cost and that the *side* of the tank will contribute *two-thirds* of the total cost, *no matter what the cost coefficients are*! The extension to more meaningful design problems of this happy principle is exciting.

4. When Does All This Work?

The formulation here is of use only if the following conditions hold:

- The number of terms minus the number of variables minus 1 is equal to 0.
- All coefficients on terms are positive.

- The constraint has but one term.
- All terms are polynomials.

Problems not conforming to these conditions *can still be solved.* Explicit applications of how to do real problems are gleefully presented in Woolsey [1991] and Beighter and Phillips [1976].

5. Sensitivity Analysis

In our example, much was made of the fact that the deltas are invariant over price or volume requirement changes. In short, no matter what the price per square yard of surface steel, the top of the tank will contribute one-third of the total cost, and the side will contribute two-thirds. The last d, the one for the constraint, now comes into play. We all know that by the time that we have designed the tank, if prices don't change, the requirements *will.* Say that we are now told that the *new* requirement is for a tank holding 1500π, rather than 1000π, cubic yards of oil. In short, if the old problem was

$$\begin{array}{ll} \text{MINIMIZE} & \pi R^2 + 2\pi RH \\ \text{subject to} & \pi R^2 H \geq 1000\pi, \end{array}$$

then the new problem has the same objective function, but now the constraint is

$$\text{subject to} \quad 1500\pi R^{-2} H^{-1} \leq 1.$$

It is important to note that a *coefficient* in the problem may have changed, but the *exponents did NOT*. This means that the three exponent equations **(4)–(6)** defined by Rule 2 before are *still*

$$\begin{array}{rcrcrcl} d_1 & + & d_2 & + & 0d_3 & = & 1, \\ 2d_1 & + & d_2 & - & 2d_3 & = & 0. \\ 0d_1 & + & d_2 & - & d_3 & = & 0, \end{array}$$

and they *still give* $d_1 = 2/3$, $d_2 = 2/3$, $d_3 = 2/3$. This means that we just plug the new coefficient into the expression for the optimal value, which you may recall from Rule 1 is

$$\$^* = \left(\frac{\pi}{d_1}\right)^{d_1} \left(\frac{2\pi}{d_2}\right)^{d_2} (1500\pi)^{d_3},$$

or

$$\$^* = \left(\frac{\pi}{1/3}\right)^{1/3} \left(\frac{2\pi}{2/3}\right)^{2/3} (1500\pi)^{2/3} = \$1,234.99.$$

Another way to look at this result is to realize that we could get the same answer by multiplying the *old* cost of \$942.48 by the ratio of the new volume to the old volume raised to the power of the constraint's d value, or

$$\$942.48 \cdot \left(\frac{1500}{1000}\right)^{2/3} = \$1,234.99.$$

The above process can be generalized into the following Change Rule of GP.

Change Rule of GP:

If we have solved a model with GP, we can find out what a requirement change in a particular component will do to our optimal cost by simple substitution into the formula

$$\text{new total cost} = \text{old cost} \times \left(\frac{\text{component's new requirement}}{\text{component's old requirement}}\right)^{\text{component's old weight}}$$

This might become a bit clearer when we realize that we got the \$942.48 cost for the *original* problem from Rule 1, or

$$\left(\frac{\pi}{1/3}\right)^{1/3}\left(\frac{2\pi}{2/3}\right)^{2/3}(1000)^{2/3} = \$942.48.$$

So when we substitute the *original* cost into the change rule, this is equivalent to substituting the above expression for \$942.48, which gives

$$\text{new total cost} \;=\; \left(\frac{\pi}{1/3}\right)^{1/3}\left(\frac{2\pi}{2/3}\right)^{2/3}(1000)^{2/3} \times \left(\frac{1500}{1000}\right)^{2/3}.$$

Cancelling the $(1000)^{2/3}$ above and below, we get the application of the change rule for this problem, or

$$\text{new total cost} \;=\; \left(\frac{\pi}{1/3}\right)^{1/3}\left(\frac{2\pi}{2/3}\right)^{2/3} \times \times (1500)^{2/3} = \$1,234.99.$$

What this means is that we *do not have to solve the problem again*, just because we have changed a price or a requirement. Or, to put this even more strongly:

Once we have solved a model of an engineering design problem with GP, WE NEVER DO IT AGAIN.

6. Plastic Batch Reactor-Mixer Problem

An electronics manufacturer prepares a plastic in a special reactor-mixer.

The capital cost for the reactor mixer is $\$316.2 \cdot S^{1/2}$.

Overhead costs and electricity expressed in terms of power requirements come to $\$34.3 \cdot P$.

Direct power may be expressed as $\$10^8 \cdot P^{-1} S^{-1/2}$.

Now, if S is the capacity of the reactor-mixer in pounds and P is the power to the stirrer in kilowatt-hours, what is the minimum-cost design for the reactor-mixer? Use the space below for your calculations.

Total cost = __________, S^* = __________, P^* = __________.

While we are still in the design stage, our friendly cost estimator tells us that our (revised) capital cost coefficient is \$320, our new overhead and electricity cost coefficient becomes \$50, and our direct power coefficient becomes $\$1 \times 10^9$. What is the new optimal solution? *Do not* re-solve the problem.

Total cost = __________, S^* = __________, P^* = __________.

7. Shell-and-Tube Heat Exchanger Problem

A total of 300 lineal feet of tubes must be installed in a shell-and-tube heat exchanger to proved required heat transfer area. Total cost of the installation includes:

- Fixed tube cost = $\$7,000$.
- Shell cost = $\$250D^{2.5} \cdot L$.
- Structural support cost = $\$200 \cdot DL$.

Spacing of the tubes is such that 20 tubes will fit into a one-square-foot sectional area of the shell. The diameter D and the length L (both in feet) must be specified to minimize the installed cost of the exchanger. The total cost is then

$$\$ = 7000 + 250D^{2.5}L + 200DL$$

And as the heat exchanger must contain at least 300 feet of tubes, the above is subject to

$$\left(\frac{\pi D^2}{4}\right) \cdot (20) \cdot L \geq 300.$$

First, find the optimal contributions to total cost of the shell and the structural support, then find the optimal cost, and finally determine the optimal diameter and length.

The problem has ____ terms − ____ variables − 1 = ____ degrees of difficulty.

The exponent matrix from Rule 2 for the problem is:

Shell percentage is ____, structural support percentage is ____.

The form of the optimal solution from Rule 1 is:

Now use Rule 3 to find the optimal values of the variables:

The values of D and L found from Rule 3 are $D =$ ____, $L =$ ____.

8. Heat Rejection System of a Steam Power Plant

The heat rejection system of a condenser of a steam power plant is to be designed for minimum project cost. The heat rejection system generates a rejection rate from the condenser of $q = 4.8 \times 10^7$ Btu/hr.

The cost of the installed cooling tower is $\$200 \cdot A^{0.6}$.

The expression for all future pumping costs is $\$5 \times 10^{-16} \cdot m^3$.

The opportunity cost of lost power production due to cooling water temperature elevation is $\$100 \cdot t$.

We now define the variables as

A = cooling tower heat transfer area (sq ft),

m = cooling water flow rate (lb/hr), and

t = cooling water temperature (°F) entering the condenser (leaving the tower).

If the temperature leaving the cooling tower is high, then the temperature of the water being recirculated back into the condenser is high, and evaporation of the cooling water in the cooling tower is significant. Since this water has been treated with chemicals to soften it (prevent scaling), there is an obvious penalty for sending hot water out of the cooling tower. The cooling tower transfers heat as $q = 10^{-5} \cdot m^{1.2} \cdot A$.

The total cost is then

$$\$ = 200 \cdot A^{0.6} + 5 \times 10^{-16} \cdot m^3 + 100 \cdot t,$$

subject to

$$4.8 \times 10^7 \leq 10^{-5} \cdot m^{1.2} \cdot t \cdot A.$$

Solve this problem with geometric programming.

$\$^* =$ __________, $A^* =$ __________, $m^* =$ __________, $t^* =$ __________.

9. Sensitivity Analysis of the Steam Power Plant Problem

In the previous problem for the heat rejection system, our cost estimators believe that the cost of the installed cooling tower will inflate from 5% to 12% ($210 to $224) and the opportunity cost will inflate from 8% to 15% ($108 to $115).

What is now our range of costs for optimal construction (before adding our profit and pack, of course) for our bid? *Do not* re-solve the problem.

Best case for inflation: cost is $ ____________.

Worst case for inflation: cost is $ ____________.

10. Waste Treatment Plant Problem

Ecker and McNamara, in a preliminary design of a waste treatment plant [1971], show that the total annual cost in thousands of dollars for the components are:

Primary clarifier	$\$19.4 \cdot x_1^{-1.47}$
Trickling filter	$\$16.8 \cdot x_2^{-1.66}$
Activated sludge following trickling filter	$\$91.5 \cdot x_3^{-0.30}$
Carbon adsorption following activated sludge	$\$120 \cdot x_4^{-0.33}$

The quantity x_i is the fraction of five-day biochemical oxygen demand (BOD5) remaining on process i, at completion. We define K to be the required effluent quality in terms of the fraction of the original BOD5 remaining after the completion of all of the processes in the design. In this case, we require that 97.1% of the influent BOD5 be removed, which would give the constraint

$$x_1 x_2 x_3 x_4 \leq 1 - 0.971 = 0.029.$$

Formulate and solve this problem using GP.

The percentage contributions to total cost for the primary clarifier, trickling filter, activated sludge, and carbon adsorption are

$d_1 =$ __________%, $d_2 =$ __________%, $d_3 =$ __________%, $d_4 =$ __________%.

The total cost of this optimal design is \$* __________.

The optimal values of the x_is are:

$x_1 =$ __________, $x_2 =$ __________, $x_3 =$ __________, $x_4 =$ __________.

11. Waste Treatment Plant Sensitivity Problems

11.1 Sensitivity Problem 1

While working on the previous waste treatment design problem by Ecker and McNamara [1971], we discover that the Environmental Protection Agency has changed its requirement, so that it now requires that we remove 98.6% of the influent BOD5. What is the new cost of the design? *Do not* re-solve the problem.

The new cost of the optimal design is $* ___________ .

11.2 Sensitivity Problem 2

Good news! The EPA has backed off to the old BOD5 requirement of 97.1%. *But* we discover that while they were changing their mind, inflation struck our labor and materials estimates. The new coefficients for the primary clarifier, trickling filter, activated sludge, and carbon adsorption are now 20.1, 17.6, 97.6, and 125.4. What is the new cost of the design? *Do not* re-solve the problem.

The new cost of the optimal design is $* ___________ .

12. Solutions to the Problems

Plastic Batch Reactor-Mixer Problem

Total cost = \$30,822.98, $S^* = 1055.889$, $P^* = 299.5356$.
Total cost = \$75,595.27, $S^* = 6201.374$, $P^* = 503.9566$.

Shell-and-Tube Heat Exchanger Problem

The problem has 3 terms − 2 variables − 1 = 0 degrees of difficulty. The exponent matrix from Rule 2 for the problem is:

$$\begin{aligned} \delta_1 + \delta_2 \qquad\quad &= 1 \\ \tfrac{5}{2}\delta_1 + \delta_2 - 2\delta_3 &= 0 \\ \delta_1 + \delta_2 + \delta_3 &= 0 \end{aligned}$$

Shell percentage is 66 2/3%, structural support percentage is 33 1/3%. The form of the optimal solution from Rule 1 is:

$$TC^* = \left(\frac{250}{2/3}\right)^{2/3} \left(\frac{200}{1/3}\right)^{1/3} \left(\frac{60}{\pi}\right) = 8,376.69.$$

The values of D and L found from Rule 3 are $D = 1.37$, $L = 10.19$.

Heat Rejection System of a Steam Power Plant

$\$^* = 28,899.57$, $A^* = 1440.24$, $m^* = 1.96 \times 10^6$, $t^* = 94.24$.

Sensitivity Analysis of the Steam Power Plant Problem

Best case for inflation: cost is \$30,430.37.
Worst case for inflation: cost is \$32,168.81.

Waste Treatment Plant Problem

The percentage contributions to total cost for the primary clarifier, trickling filter, activated sludge, and carbon adsorption are
$d_1 = 8.9\%$, $d_2 = 7.88\%$, $d_3 = 43.6\%$, $d_4 = 39.62\%$.

The total cost of this optimal design is \$387,881.20.

The optimal values of the x_is are:
$x_1 = 0.6687$, $x_2 = 0.6909$, $x_3 = 0.1227$, $x_4 = 0.4508$.

Waste Treatment Plant Sensitivity Problems

Sensitivity Problem 1: The new cost of the design is \$426,707.30.
Sensitivity Problem 2: The new cost of the design is \$408,749.30.

References

Duffin, R.J., E. Peterson, and C. Zener. 1967. *Geometric Programming.* New York: Wiley.

Beighter, Charles, and Don T. Phillips. 1976. *Applied Geometric Programming.* New York: Wiley.

Ecker, J.G., and J.R. McNamara. 1971. GP and the preliminary design of industrial waste treatment plants. *Water Resources Research* 7 (1) (February 1971): 18–22.

Woolsey, R.E.D. 1991. *Engineering Design and Optimization with Geometric Programming.* Golden, CO: Colorado School of Mines.

About the Author

Robert E.D. ("Gene") Woolsey received B.A., M.A., and Ph.D. degrees in mechanical engineering from the University of Texas at Austin; he is professor of mineral economics at the Colorado School of Mines. He is a past president of the Institute of Management Sciences and a past editor of its journal *Interfaces*, for which he still write an occasional "Fifth Column" of war stories. Prof. Woolsey refers to himself as the "principal gadfly, to the operations research and management science community, for more practical utility and fun in the profession." His principal aim in life is to graduate the entrepeneurs of tomorrow so that the socialists will have someone to tax.

UMAP

Modules in Undergraduate Mathematics and Its Applications

Published in cooperation with the Society for Industrial and Applied Mathematics, the Mathematical Association of America, the National Council of Teachers of Mathematics, the American Mathematical Association of Two-Year Colleges, The Institute of Management Sciences, and the American Statistical Association.

Module 739

Measuring Stream Discharge: A Precalculus Field Exercise

Steven Kolmes
D. Brooks McKinney
Kevin Mitchell

Applications of Environmental Sciences to Precalculus

COMAP, Inc., Suite 210, 57 Bedford Street, Lexington, MA 02173 (617) 862–7878

INTERMODULAR DESCRIPTION SHEET: UMAP Unit 739

TITLE: Measuring Stream Discharge: A Precalculus Field Exercise

AUTHORS:
Steven Kolmes
Dept. of Biology
University of Portland
Portland, OR 97203

D. Brooks McKinney
Dept. of Geoscience

Kevin Mitchell
Dept. of Mathematics and Computer Science

Hobart and William Smith Colleges
Geneva, NY 14456

MATHEMATICAL FIELD: Precalculus

APPLICATION FIELD: Environmental and earth sciences

TARGET AUDIENCE: Students in a course in precalculus.

ABSTRACT: This Module consists of a set of four collaborative field-based exercises and labs. All of the materials derive from efforts to estimate, measure, and model the flow of a local stream. During the course of these four sections, students work with quadratic equations to model the stream flow; in making their discharge estimates, they are introduced to Riemann sums, a central concept that they will encounter in later calculus courses.

PREREQUISITES: Familiarity with quadratic equations and their graphs.

COMAP, Inc., Suite 210, 57 Bedford Street, Lexington, MA 02173
(800) 77–COMAP = (800) 772–6627 (617) 862–7878

Measuring Stream Discharge: A Precalculus Field Exercise

Steven Kolmes
Dept. of Biology
University of Portland
Portland, OR 97203
`kolmes@uofport.edu`

D. Brooks McKinney
Dept. of Geoscience
Hobart and William Smith Colleges
Geneva, NY 14456
`dbmck@hws.bitnet`

Kevin Mitchell
Dept. of Mathematics and Computer Science
Hobart and William Smith Colleges
Geneva, NY 14456
`mitchell@hws.bitnet`

Table of Contents

MODULES AND MONOGRAPHS IN UNDERGRADUATE MATHEMATICS AND ITS APPLICATIONS (UMAP) PROJECT

The goal of UMAP is to develop, through a community of users and developers, a system of instructional modules in undergraduate mathematics and its applications, to be used to supplement existing courses and from which complete courses may eventually be built.

The Project was guided by a National Advisory Board of mathematicians, scientists, and educators. UMAP was funded by a grant from the National Science Foundation and is now supported by the Consortium for Mathematics and Its Applications (COMAP), Inc., a nonprofit corporation engaged in research and development in mathematics education.

Paul J. Campbell	Editor
Solomon Garfunkel	Executive Director, COMAP

1. Project Introduction: Water, Water Everywhere

1.1 The Hydrologic Cycle

Water is of critical importance to all living things. Not only do all life forms require water, but the availability of water is a critical factor in climate as well as ecosystem structure. The amount of water on Earth is essentially fixed, but this water is seldom static. Water constantly cycles through the ocean-atmosphere-terrestrial system in what geologists refer to as the *hydrologic cycle* (**Figure 1**).

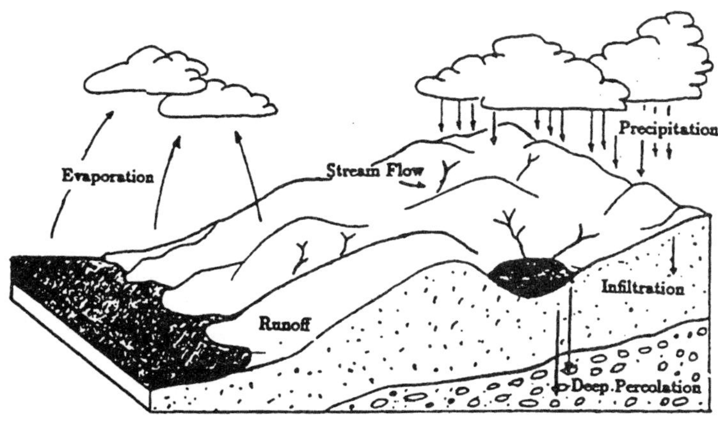

Figure 1. The hydrologic cycle (U.S. Dept. of Agriculture, as adapted from Figure 4–12 of Miller [1988]).

Stream flow is one component of this cycle. The water for stream flow typically begins the cycle by evaporation into the atmosphere, transportation by weather systems, and precipitation as rain or snow onto the land surface. Once on the surface, this water may re-evaporate (perhaps after uptake by plants, transpiration), infiltrate into the ground (groundwater), or flow overland under the influence of gravity as runoff. Small runoff tributaries join together to create streams. The land area drained by a stream and its tributaries is that stream's *watershed* or drainage basin. As **Figure 1** indicates, there are many different paths that water can follow through the cycle. For example, groundwater also moves under the influence of gravity, infiltrating into the ground in one area and reappearing at the surface as springs or seeps down slope. For the purposes of this set of exercises, though, it is only important to know that the rate of stream flow, runoff, is a function of the amount of precipitation that falls on a watershed and the size of the watershed. Because some of this precipitation re-evaporates, the annual volume of runoff from a basin is normally less than the annual volume of precipitation that falls upon it.

1.2 Watersheds

The concept of the watershed is critical to the study of streams. The watershed is the fundamental "unit" for consideration of stream flow. Typically, precipitation falling on the stream's watershed supplies nearly all of the water in the stream. Hence, to a first approximation, stream flow and stream characteristics are functions only of what happens in a particular watershed. This idea has important environmental implications, since pollutants introduced into an area may pollute surface waters in that watershed but not in other watersheds.

Because water flows downhill, the area of a watershed is determined by the shape of the regional land surface, the topography. The boundary of a watershed is called a *drainage divide*. A drainage divide is a line that separates one watershed or drainage basin from another. Physically, drainage divides are local topographic highs from which the land surface slopes downward in two or more distinct directions. The classic example of a drainage divide is a sharp mountain ridge separating two valleys. The crest of the ridge is a drainage divide, dividing falling precipitation and consequent runoff into one of the two valleys, depending upon which side of the ridge the precipitation falls.

Drainage divides can be located using a topographic map on which elevations are represented as contours of equal elevation. **Figure 2a** is a contour map of a small area in northern New York. For any pair of contour lines on this map, one line will be higher than the other.

The direction of maximum downward slope, and so the direction of runoff flow, is a line orthogonal to the two adjacent contour lines and pointing away from the higher contour and toward the lower contour (**Figure 2a**). Drainage divides are lines along which these slope directions diverge (**Figure 2a**). (Streams or stream valleys lie along lines of slope convergence.) Once the basic "downhill" principle illustrated in **Figure 2** is understood, it is seldom necessary to draw downhill arrows on a topographic map to identify drainage divides. Instead, slope directions can be estimated by eye as the divide is drawn.

There are a few "rules of thumb" that are helpful in drawing drainage divides. First, drainage divides always form a closed loop. If you begin drawing a divide from some point along a stream, it necessarily will encircle the upper portion of that stream and all of its upstream tributaries, finally rejoining the stream at the starting point. All upstream tributaries to the stream must be within this loop, all downstream tributaries or other streams must be outside of the loop. The drainage divide cannot cross any flowing stream (except the one stream for which it is constructed, and then only at one place, the outflow of the basin).

These rules allow the approximate location of drainage basins even on maps lacking topographic information. **Figure 3** is a map showing only the map patterns of streams in the northeastern United States. In spite of that, an approximate drainage basin for the Susquehanna River can be drawn just by locating the divide between tributaries of different main streams. In practice, even with topographic information, these rough rules are used to approximately locate drainage divides and then a more detailed analysis of contour lines and slopes

Figure 2a. A contour map of a small area in northern New York. Arrows show diverging down-slope directions orthogonal to contours. Drainage divides must lie between such diverging directions. **b.** The same map with the drainage divide drawn in (dotted line). Note that it is a closed loop and does not cross any other streams.

1.3 Discharge

The volume of stream flow per unit time passing a particular point along a stream is that stream's *discharge* at that point and time. Imagine a stream of water coming out of your kitchen tap. If you were to put a container under the tap and catch all the water coming out for some unit of time (day, minute, second) and measure the resulting volume of water, that would be the discharge of your tap. Discharge is always measured in units of volume per unit time. Typical discharge units are cubic feet per second or cubic meters per second. The discharge for a stream generally increases downstream as tributaries are added. It also varies temporally with seasonal precipitation and snow melt.

The total annual discharge from a watershed will depend upon the volume of water that falls as precipitation on that watershed, and how much of that water is lost to evaporation, transpiration, groundwater storage, etc. Typically, annual discharges from temperate climate basins are about half of the volume of annual precipitation.

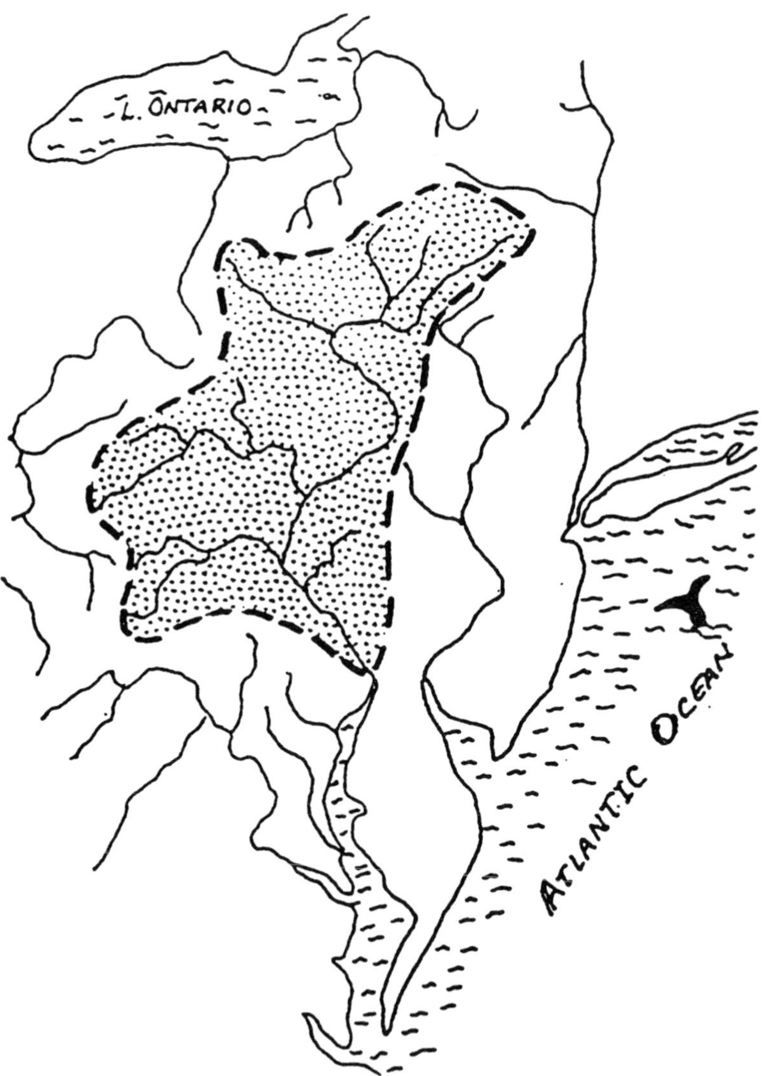

Figure 3. Map patterns of streams in the northeastern United States. The approximate drainage divide (dashed line) of the Susquehanna River watershed (patterned area) can be drawn as a loop that surrounds all the Susquehanna's tributaries but does not cross any other streams.

Example. Consider the following situation. Suppose a small stream has a watershed of 5 km^2 (square kilometers). Further, assume that the rainfall on the watershed is 1.5 meters per year (m/yr). Then an estimate for the maximum volume of water to pass out of the stream in a year should be equal to the volume of a depth 1.5 m spread over 5 km^2. Let's convert this volume to cubic meters (m^3). A kilometer is 1,000 m. So a square kilometer is $(1{,}000 \text{ m})^2$ or 1,000,000 m^2. Therefore the discharge over the course of a year would be:

$$\begin{aligned} Q &= 1.5 \text{ m/yr} \times 5 \text{ km}^2 \\ &= 1.5 \text{ m/yr} \times 5 \cdot 1,000,000 \text{ m}^2 \\ &= 7,500,000 \text{ m}^3/\text{yr}. \end{aligned}$$

To convert this annual discharge into cubic meters per day, first change years into days by using the conversion factor

$$\frac{1 \text{ yr}}{365 \text{ days}}.$$

So now

$$Q = 7,500,000 \text{ m}^3/\text{yr} \times \frac{1 \text{ yr}}{365 \text{ days}} \approx 21,000 \text{ m}^3/\text{day}.$$

This value is the *average* daily discharge based on an annual discharge. Using further conversions, we could compute a discharge rate in units of cubic meters per second (m^3/s). These discharges are *maximum* values. For temperate climates, one would expect *actual* discharges to be about 50% of these values.

2. Lab 1: The Water Falling on a Watershed

Areas of regular shapes, such as rectangles or triangles, are easy to compute, using formulas involving base and height. Irregularly shaped objects have areas that can be difficult to compute. In fact, much of first-year calculus is spent finding ways to compute the areas of irregular shapes.

Watersheds generally have very irregular shapes (see **Figures 2** and **3**), so we have to be clever to compute the area of a watershed. Three different methods to compute such areas are outlined in the exercises. The first method involves tracing the watershed onto a piece of graph paper, where each square of the graph paper represents a known area. The second method involves weighing a cut-out of the watershed and comparing the result to the weight of a piece of paper representing a known area. The third method involves using a computer and a digitizing tablet. All require being able to convert units of measurement accurately, so be careful!

2.1 Map Analysis

Each pair of students will be given a photocopy of a map with the watershed outlined and sheets of graph paper. Each pair should carefully write up the answers to the following questions.

Exercises

1. **The Map Scale.** Find the linear scale on your map.
   a) How many meters does 1 centimeter on your map represent?
   b) How many kilometers does 1 centimeter on the map represent?
   c) How many square kilometers does 1 square centimeter on the map represent? You may wish to express your answer as a conversion factor of the form (field km^2)/(map cm^2).
   d) Determine the number of square centimeters in one square of the coarse grid on the graph paper. What area does this represent on the map?
   e) Now determine the area of one square of the fine grid on the graph paper and the corresponding area on the map.

2. Take your copy of the map and carefully cut out the watershed with a pair of scissors. This cutout can be used to measure the watershed in three different ways.

3. **Using a Grid.** One method to approximate the area is to outline the boundary of the watershed on graph paper.
   a) Using a sharp pencil, make such an outline on the graph paper supplied.
   b) Using the coarse grid on the paper, count the squares that are *entirely* enclosed by the drainage divide and ignore any others.
   c) Use your conversion factor and the number of squares counted to approximate the area of the watershed. Have you overestimated or underestimated the area of the watershed in using this process? Explain.
   d) Now using the coarse grid on the paper, count all the squares that even partially overlap the watershed.
   e) Use your conversion factor and the number of squares counted to approximate the area of the watershed. Have you overestimated or underestimated the area of the watershed this time?
   f) Can you give an upper and lower bound on the size of the watershed?
   g) Repeat these same two processes, but use the fine grid on the graph paper this time. (You can avoid counting many of the individual fine squares by counting them a coarse grid block at a time.)
   h) Can you revise the upper and lower bounds on the size of the watershed?
   i) Compare your values to others in the class. Assuming that you all traced and counted carefully, why might you still get different answers?

4. **Weighing the Watershed.** Another way of finding the area is to weigh it!
   a) Using a balance, carefully determine the weight (in grams) of a sheet of graph paper that has had its margins removed. (Note: For accuracy, make sure that the entire piece of paper is contained in the balance pan. If the paper hangs over the edge, fold it.)
   b) What size area would this rectangular region represent on the map?
   c) Use your two preceding answers to determine a conversion factor in units of the form km^2/gm. You now have a way to convert weight to area!
   d) Weigh your cut-out of the watershed (fold it if necessary). Convert this weight into an approximate area of the watershed.

5. **Digitizing.** A digitizing tablet can also be used to determine the area of the watershed. In this method, an electronic pen or mouse is connected to a computer. When the boundary of a figure is traced with the pen or mouse, the computer calculates the area in square centimeters.
   a) Trace your watershed boundary on the digitizing tablet. How many square centimeters did it enclose?
   b) Use a conversion factor from **Exercise 1** to determine an approximation of the area of the watershed.

6. Compare all of the area approximations that you have obtained for the watershed. Do the approximations from weighing and/or digitizing lie between the upper and lower bounds established by the grid analysis? Which value do you feel is most accurate? Least accurate? Explain.

7. **a)** Using the average annual (month's) rainfall for the watershed and what you think is your most accurate estimate for the watershed area, calculate the volume of water falling on the watershed per year (month).
   b) Convert this rate to units of m^3/sec.
   c) What would the discharge for the stream be (in m^3/sec) if all the rain that fell on the watershed reached the stream?

3. Lab 2: Measuring Stream Discharge

Rivers and streams are dynamic systems. Anyone familiar with a particular stream will have noticed that the depth and velocity of a stream vary by season and in response to periods of very wet or dry weather. Changes in stream depth and velocity reflect temporal changes in stream discharge. The purpose of this lab is to estimate the discharge of a stream at one particular level of flow.

3.1 Background and Assumptions

To simplify the field work and later modeling, it is useful to assume that the stream and its channel have certain characteristics.

- The stream should be straight where it is being measured.
- The banks of the stream should not be undercut by the stream.
- The stream should be flowing through dry terrain, not through a wetland.

With these assumptions, the theory behind the measurement of discharge is relatively simple. The amount of water flowing past a given point per unit time is the discharge, Q. By definition, this is equal to the cross-sectional area of the river multiplied by the velocity. That is,

$$Q = A \cdot V,$$

where Q is the discharge rate, A is the cross-sectional area of stream, and V is the average velocity of the stream. If the section of stream were nearly rectangular, the cross-sectional area could be found by multiplying the width by the depth of the stream: $A = w \cdot d$. Substituting into the previous equation gives the discharge as:

$$Q = w \cdot d \cdot V.$$

Let's try to picture what happens to width and depth for normal flows as discharge increases, for example, as a response to increased rainfall runoff. One can imagine that the water will have to run faster, or the width or depth will have to increase.

There are two simple observations about the flow rate that we can make based on our "everyday" experiences with rivers. First, a river flows the fastest along its center, and the velocity is almost 0 near each bank. You may have noticed this if you have floated leaves or sticks in a stream or river. Those closest to the center travel the fastest downstream. Or, if you have paddled a boat downstream, you tend to go faster when you are in the middle of the stream.

The second observation is that, if the stream is straight, the flow rate depends only on the distance from the center of the stream, not on whether it is the right- or left-half of the stream. That is, the velocity of the water is *symmetric* about the center line of the stream. Near either edge of the stream, the velocity is almost 0; and it increases as one approaches the center.

3.2 Methods

The velocity of the stream varies as we move across it. However, we can imagine that the stream is divided up into sufficiently small sections over which the velocity is nearly constant, as illustrated in **Figure 4**.

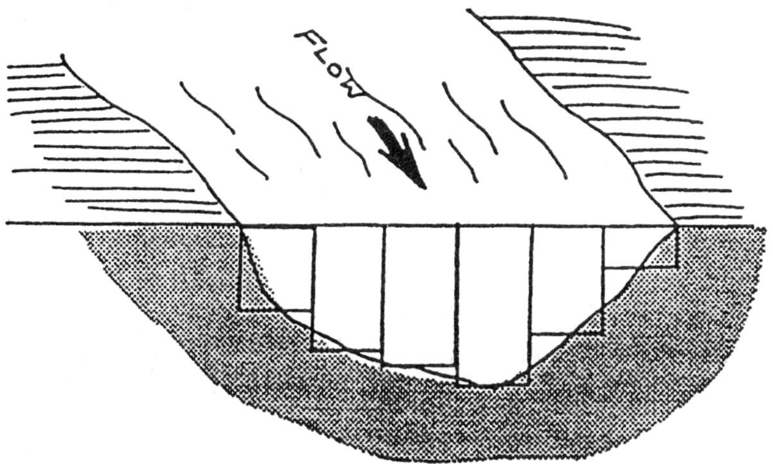

Figure 4. A subdivision or partition of a stream into equal-width sections.

For each section, we approximate its cross-sectional area as a rectangle, with height equal to the stream depth in the center of the section, and width equal to the width of the section.[1] If this area is multiplied by the stream's velocity, the result is discharge, the volume of water that moves through that cross-sectional area in unit time. That is, $A \cdot V = Q$.

This is the discharge for one particular section. Because adjacent sections will have different depths and velocities, each section will likely have its own unique discharge. The discharge of the entire stream is the sum of the discharges for each section.

The measurements of the stream width and depth can be made directly, using a tape measure and a graduated wading staff. To measure the stream's velocity, we will use a stopwatch to measure the time required for a neutral buoyancy float to travel one meter downstream. The reciprocal of this travel time is the velocity in meters per second. (Do you see why?)

3.3 Materials and Preparation

You will be working in the water and mud along a stream bank, so wear old clothes and shoes that you don't mind getting wet or muddy. The measurements will be done in groups of four; one person in the group will be outfitted

[1]This idea of subdivision of a region into rectangles is crucial in calculus. We used a similar process to find the area of the watershed.

with a pair of hip boots (if needed) to wade across the stream. If others want to get in the stream, they should wear old pants or shorts and sneakers. Barefoot wading is dangerous, because the bottom may contain broken glass or fish-hooks mixed in the mud.

Before leaving, each group should check to make sure that they have the following items:

- 30 m tape measure,
- calculator,
- stopwatch,
- graduated wading staff (or meter stick),
- neutral buoyancy float (a small piece of sponge or a Ping-Pong ball with two holes),
- Stream Discharge Worksheet,
- pencils, and
- hip boots (if needed).

3.4 Instructions for Data Collection and Analysis

Upon arriving at the stream, each group will be given a specific location along the bank to make their measurements. At your group's location, carry out the following steps:

1. Have one person who agrees to carry out the depth and velocity measurements put on hip boots (if needed). Have a second person roll up their pants but leave their sneakers on to protect their feet.
2. The person with rolled-up pants should wade across the stream, carrying one end of the 30 m tape measure. Making sure that it is being held tightly by another team member, measure the width of the water at your location. Remember that if the tape measure is held slackly, you won't get an accurate width measurement. Fill the answers to **Exercise 1** on the **Stream Discharge Worksheet**.
3. Divide the stream into ten equal-width zones and record the midpoint of each zone under the heading "Station Location" on your stream discharge worksheet. For example, a 10-meter-wide stream would divide into ten 1-meter-wide bands, the midpoints of which would be 0.5 m, 1.5 m, 2.5 m, 3.5 m, 4.5 m, 5.5 m, 6.5 m, 7.5 m 8.5 m, and 9.5 m. **Remember:** start at the midpoint of your first band for your first measurement, and then move by increments equal to your width Δw for a single band. All too often, students mistakenly start at the far edge of the first band, and end up having to measure the entire stream's flow over again.

4. For each station's midpoint, have the person in hip boots use the graduated wading staff to measure the depth. Record these values in the appropriate column on the **Worksheet**.

5. Next, your group needs to measure how long it takes a neutral buoyancy float (either a bit of sponge or a Ping-Pong ball) to float 1 m along the surface of the water at each station's midpoint. This can be done by having the hip-boot-wearer find the station midpoint again (which requires that two other people are holding the tape measure tightly) and use the graduated wading staff to drop a float 1 m upstream of the tape measure. She/he should shout "Go!" when the float is dropped and "Stop!" when it has traveled 1 m and crossed under the tape measure. Another team member should use the stopwatch to time how many seconds it takes the float to travel 1 m. Note that you'll need to recover the float, which may require having a team member (the person with the stopwatch) stand downstream of the float to catch it. This procedure therefore requires four people to cooperate. Record in the "Time" column on the **Worksheet** the time that it takes to float 1 m. If time permits, repeat the time trials at each station three to five times and record the *average* time that it takes the float to travel 1 m.

6. Once you have made all ten measurements, have your group members rejoin on the bank and make sure that you have all your equipment. Make sure that each group member has a copy of all data.

7. With the data that you have collected, you can compute the Total Discharge of the stream.
   a) Begin by using the values in the "Time" column to determine the values in the "Velocity" column on the worksheet. Since velocity is distance divided by time, and since the distance the float has traveled is one meter, we have: $V = 1/t$. That is, in this experiment, velocity is just the reciprocal of time.
   b) Next calculate a "Discharge" value for each station: $Q = \Delta w \cdot D \cdot V$. (Your discharge units should be m^3/s.)
   c) Calculate the "Total Stream Discharge," by adding the discharges from all ten stations, and enter this value on your worksheet.

8. Compare your values with a partner from your field group. You should get the same results. Then compare results with different groups; their results should be similar, if they were located close enough to your group.

9. a) Compare the actual discharge for the stream to the theoretical discharge that you calculated in the earlier lab. Use units of m^3/s.
   b) If the theoretical and actual discharges are different, give two hypotheses that might be the explanation.

3.5 Stream Discharge Worksheet

1. Measure the width w of the stream with the measuring tape and then determine each of the ten station positions.

   (a) The width of the stream,

   $$w = \_\_\_\_\_\_\_\_ .$$

   (b) The width of the ten subdivisions of the stream,

   $$\Delta w = \frac{w}{10} = \_\_\_\_\_\_\_\_ .$$

2. Fill in the first three columns of this table as you make your measurements. The rest of the table can be completed after all measurements are complete.

	Station Location	Depth D, m	Time t, sec	Velocity V	Discharge Q, m^3/sec
1					
2					
3					
4					
5					
6					
7					
8					
9					
10					

3. After filling in the entire table above, calculate the Total Stream Discharge:

   $$Q = \_\_\_\_\_\_\_\_ .$$

4. Simple Models of Stream Discharge

Bolt [1984] suggests several simple models of stream velocity. We consider two such models here. These models use the two observations that we made earlier about stream velocity:

- the velocity is roughly symmetric about the center of the stream;
- the velocity is greatest at the center and decreases to 0 toward the banks.

4.1 A Linear Model

The simplest model would have the velocity of the stream decrease linearly from the maximum value, say M m/s at the center of the stream, to 0 m/s at the edges of the stream. This is illustrated in **Figure 5**, where we have assumed that the stream has total width w.

Using the three points marked in the graph, it is straightforward to calculate the equations of the two line segments that make up the graph of the stream's velocity. We get

$$v(x) = \begin{cases} \dfrac{2Mx}{w}, & \text{if } 0 \le x \le w/2; \\[2ex] \dfrac{-2Mx}{w} + 2M, & \text{if } w/2 < x \le w. \end{cases}$$

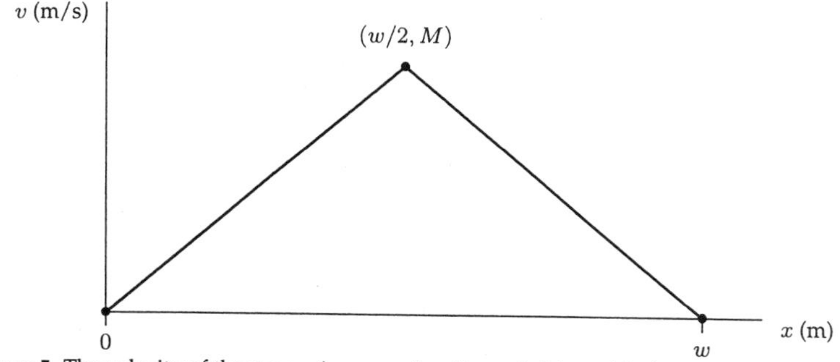

Figure 5. The velocity of the stream increases in a linear fashion with distance from the bank.

4.2 A Quadratic Model

Based on actual experience, there is reason to believe that the flow rate of the river does not drop off so rapidly as in **Figure 5**. (It is actually friction near the banks that causes the velocity to decrease.) Rather, the graph of stream velocity is more like the graph of an upside-down parabola, as in **Figure 6**.

The equation of the graph of the quadratic in **Figure 6** can be found as follows. Since the velocity is 0 at both edges of the stream, the velocity function

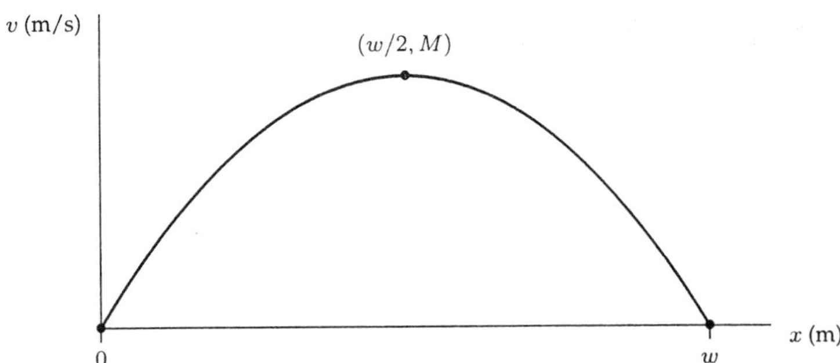

Figure 6. Another possible graph of stream velocity: a quadratic.

has roots at $x = 0$ and $x = w$. This means that the quadratic velocity function must factor as $v(x) = cx(x - w)$, where c is a constant. To evaluate c, we use the third point that we know on the parabola: At the center of the stream ($x = w/2$) the velocity has its maximum value M m/s. That is, $v(w/2) = M$. But then

$$v\left(w\right) = c \cdot \frac{w}{2} \cdot \left(\frac{w}{2} - w\right) = M,$$

or

$$-\frac{cw^2}{4} = M.$$

That is,

$$c = -\frac{4M}{w^2}.$$

Thus, we conclude that

$$v(x) = -\frac{4M}{w^2}x(x - w).$$

Figure 7 shows a graph of stream velocity and its quadratic approximation, based on data collected by students in our class.

Exercises

1. Assume that a stream is 12 m wide and that it has a maximum velocity of 3 m/s at the center.
   - **a)** Under the hypotheses of the linear model, what is the equation that describes the velocity for the stream?
   - **b)** Under the hypotheses of the quadratic model, what is the equation that describes the velocity for the stream?
2. **a)** Use the data that you collected to model stream discharge with each of the two models outlined above.
   - **b)** Graph each of these equations on the same set of axes.
   - **c)** Using the data that you collected, draw a graph of actual stream velocity on the same axes as in the previous part. Do either of the models approximate the actual data very well?

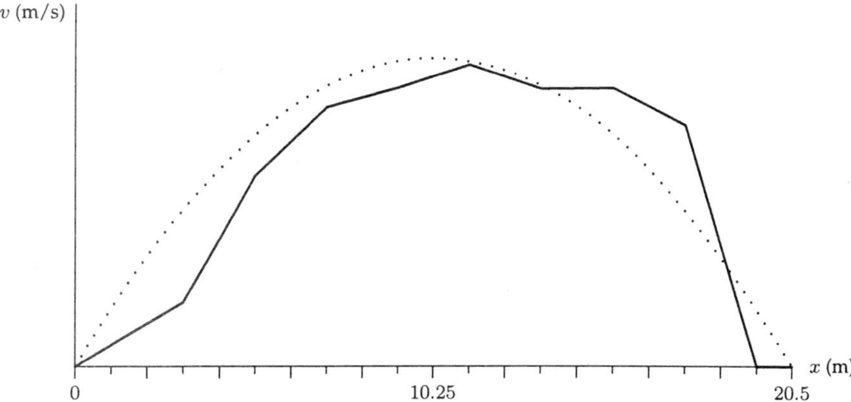

Figure 7. A graph of stream velocity (solid line) and its quadratic approximation (dotted line).

5. Appendix: Instructional Overview

5.1 Introduction

Many recent "calculus reform" efforts include projects in which students work collaboratively in small groups and report their results back to the class or instructor. "Field" projects requiring students to collect their own data, as well as to work with it, yield particularly enthusiastic feedback from students and correspondingly good work [Mitchell 1993]. The field component of these projects has many benefits. Most students enjoy the "hands-on" data collection process and develop from it an intuitive sense of what their measurements represent. For many students, this experience is the first time in 13 years of mathematics courses they have measured and worked with their own data! Later, when students work with these data, they are "vested" in it and interested in the project's outcome in a way not otherwise possible. Students also enjoy working in small groups and from this develop a camaraderie not typically present in mathematics classes.

Less attention has been paid to the precalculus training of students, but it is especially important to develop interesting materials for these students to encourage them to continue their mathematical education. Collaborative "field" projects are particularly well suited to this task.

This Module presents a set of four collaborative field-based exercises and labs that we used in a five-week summer precalculus program to help "academically at risk" incoming first-year students adjust to the demands of college while improving their quantitative abilities. All of the materials derive from efforts to estimate, measure, and model the flow of a local stream. During the course of these four sections, students work with quadratic equations to model the stream flow; in making their discharge estimates, they are introduced to Riemann sums, a central concept that they will encounter in later calculus courses.

In the rest of this Appendix, we present an overview of the pedagogy of the project, a description of the material and equipment requirements for the labs, and practical suggestions for carrying out the project.

5.2 Pedagogical Issues

Our primary goal was to take mathematical concepts familiar to entering college students (e.g., average rates of change, quadratics, area, and volume) and apply them to a "hands-on" project involving data collection and estimation, issues that are of vital importance in all of the natural sciences. The application of these ideas in a new context can serve as a much needed review of these notions almost without the students noticing it. Additionally, during the course of the project, we introduce students to new mathematical concepts that will be central to later courses in mathematics. In particular, the notion of a Riemann sum surfaces quite naturally twice in this unit, though we never introduce summation notation or limits.

Much of this project was carried out by students working in small groups (pairs or pairs of pairs). This seems particularly effective in working with students at this level. For example, students feel much more comfortable asking questions of the instructor as part of a group, and an instructor can be much more focused in interacting with a particular group than with an entire class, while still maintaining an efficiency that individual attention does not permit. Small group exercises lead to personal interaction with students while not putting any student "on the spot." One of the most satisfying outcomes of small group work is that students teach each other in this setting. We often found students reinterpreting our instructions or directions for each other, or helping each other through complicated calculations. Lab or field exercises that require all group members to cooperate for the entire group to be successful are effective in preventing one student from dominating or another from not contributing. (For an excellent introduction to the pedagogical issues related to small group learning, see Weissglass [1993].)

5.3 Project Introduction

The "Project Introduction" (**Section 1**) provides students with the background that they will need to do the project. Important concepts presented in this section are the hydrologic cycle, watersheds, drainage divides, and stream discharge. The materials in this section can be assigned to be read either in preparation for, or as review of, an introductory lecture explaining the project. Though there are no exercises in this section, the example discharge calculation is crucial to the following lab exercises. Note that the units of measurement for stream discharge, volume per unit time, are important. In the example, discharge is calculated as a volume (depth of rainfall times watershed area) per unit time (year, month, or second). Later, in the field, discharge is calculated differently, as velocity times a cross-sectional area.

The process of determining the boundary of a watershed (which is called a *drainage divide*) is described in some detail. We suggest that the instructor give an in-class demonstration of how to draw a drainage divide, using an overhead of a topographic map, but that in the subsequent labs and exercises, students work with copies of maps on which the drainage divides have already been drawn. There is typically some ambiguity in drawing drainage divides. Providing students with the drainage divides removes this ambiguity from later exercises, where they are comparing different estimation techniques. If time permits and the audience is appropriate, then topographic maps can be used to introduce the notion of level curves and gradient vectors, important topics in second-year calculus.

Ideally, the demonstration of how to draw a drainage divide and subsequent map-based calculations should be done on a topographic map that covers the watershed of a local stream where the students can measure discharge (see **Section 3**). In this case, the drainage-divide loop will start on either side of the stream at the location where students will measure discharge.

Topographic maps are available in most college or university libraries and in many book and outdoor stores. Geology, geoscience, or geography departments will commonly have maps to loan out or ordering information for local maps. Topographic maps are available in several standard scales. The best scale to use is determined by the size of the watershed. For our project, we used a map with a scale of 1:250,000 because the map representation of the watershed fit conveniently on a standard 8.5 × 11 sheet of paper but was not so small as to be difficult to measure.

5.4 Lab 1

The goal of this lab (**Section 2**) is to estimate the discharge for a watershed, using rainfall data and watershed area. Sources of rainfall data include rainfall maps in atlases, tabulations of climatological data published by the National Oceanic and Atmospheric Administration (NOAA), county agricultural extension agents, and a variety of almanacs. Requiring students to find and document a source of these data is a useful exercise in itself. An hour in practically any library will uncover several different values, reflecting particular years, averages over different periods of years, etc. This can lead to a productive discussion of the uncertainty of natural processes and which value is the "right" one.

We recommend that estimates of the area of the watershed be based upon a topographic map with the drainage divides already drawn in. Typically, the watershed will have an irregular shape; and much of this lab is concerned with estimating its area. Three different methods are used.

The first method is essentially a Riemann-sum process with lower and upper sums. Students trace the drainage divide onto a sheet of graph paper. To facilitate class comparisons, we suggest that all students be given paper with the same ruling. A particularly convenient type is ruled in bold at one-centimeter intervals, with each centimeter subdivided into fifths with a lighter ruling. The students then count the number of squares totally enclosed by the watershed and the minimum number of squares required to completely enclose the divide. Using the map scale, students convert this information into square kilometers, producing upper and lower bounds (sums) for the area of the watershed. They repeat the same process with the lighter ruling. They should notice that the upper bound decreases and the lower bound increases as the grid gets finer. (This observation is critical when it comes time to learn about integration in a calculus course.) Students should be urged to think about a process that would further improve their area estimate.

With a map whose scale was 1:250,000 and for our oblong watershed with its lengthy boundary, the smaller grid yielded lower and upper area estimates of 579 and 649 km^2. Note that repositioning the grid on the graph paper will cause a fair amount of variation in estimates, especially with the coarser grid. This exercise can be assigned as homework, if necessary.

The second method of measuring the area of the watershed requires scissors and a balance capable of measuring hundredths of a gram. Such balances are available in a biology or chemistry department. Students weigh a cut-out of their watershed and compare it to the weight of a piece of paper of known area. (We had students compare it to another sheet of graph paper with the margins removed.) Make sure that the paper is of the same type and that the entire piece of paper lies on the pan of the balance (fold if necessary). Using this method, all students obtained an area estimate that was between the lower and upper values obtained using their fine grid.

The third method requires a digitizing tablet and its accompanying software. Look to a biology or geology department for such equipment. You will need to explain how to use your particular tablet. Depending on the software, you can have the computer or the students do the conversion to square kilometers. What we found interesting was that without too much thought, students generally picked the answer from this method as the most accurate. In fact though, it was the least accurate. Because of the difficulty of tracing an irregular shape with a mouse, most groups got an estimate that was outside the range of possible values as determined by the small-grid method. This provides an opportunity to talk about technology, estimation, and the power of Riemann sums.

5.5 Lab 2

This lab is a field exercise in which students measure stream discharge. As described in **Section 3**, this is accomplished by dividing the stream's width into sections of equal length and for each section estimating cross-sectional area (section length times depth) and velocity. Velocity is estimated by measuring the time for a neutral buoyancy float in the center of the section to move 1 meter downstream. The discharge of each section (area times velocity) is then computed, and the sections added as a Riemann sum.

Most of the equipment for the project—graduated wading staffs or meter sticks, tape measures, stopwatches, neutral buoyancy floats (a small piece of sponge or a ping-pong ball with two holes), and hip boots (if needed)—can be borrowed from your colleagues in biology or geology. Doing so will generate lots of questions on their part and perhaps some useful suggestions.

Probably the most important piece of "equipment" is the stream itself. To make their measurements, students must wade across the stream; so safety is the paramount issue in selecting an appropriate stream. The ideal stream is 8–20 m wide, no more than about 1 m deep, with a flow velocity that is easily discernible but not fast enough to be difficult to wade. Very narrow or very shallow streams are not well suited for measurement, because of the large effect that bottom and bank irregularities will have on the measurements. Larger or more swiftly flowing streams are dangerous. Instructors should be aware that flow conditions can change rapidly. The stream that may be ideal for measurement on one day may be dangerous and impossible to use the next. As in all field work, common sense and a back-up plan are essential. We have two streams that we regularly use for discharge measurements—a larger one for low-flow conditions, and a smaller but safer one for high-flow conditions. Once again, colleagues in geology or biology can probably offer suggestions for appropriate stream measurement sites.

Detailed instructions for the data collection appear in the unit itself. However, expect some initial confusion at the stream site, no matter how much preparation you have done (e.g., the student with the tape measure may wade across the stream to measure it only to realize that he or she has not given one end of the tape to another member of the group on the initial bank of the stream). However, students tend to quickly settle down and actually accomplish their work in fairly good time. Since all group members are involved, this builds camaraderie. This lab turned out to be the students' favorite part of our course. Why should the biologists and geologists have all the fun of going on field trips?

The data analysis (**Exercises 7–9** in this section) is straightforward and can be done as homework. Expect to get actual discharges that are about 50% of the estimate made in the earlier lab. Students should be able to provide reasons why this is the case.

In the "Models of Stream Discharge" section, **Section 4**, we sketch two simple mathematical models of stream flow. The models are an opportunity to review equations of lines and quadratics. Bolt [1984] provides a few more such models and is quite accessible. One could also talk about other sorts of flow at this point. (We discussed laminar blood flow in an artery. There the velocity of the "stream" is modeled by the expression $k(R^2 - r^2)$, where R is the radius of the artery and r is the distance from the center of the artery. This model is similar to the quadratic model of stream flow that is discussed in this Module.)

References

Bolt, Brian. 1984. Modeling river flow. *The UMAP Journal* 5 (2): 133–140.

Miller, G. Tyler, Jr. 1988. *Living in the Environment.* 5th ed. Belmont, CA: Wadsworth.

Mitchell, Kevin. 1993. Calculus in a movie theater. UMAP Modules in Undergraduate Mathematics and Its Applications: Module 729. *The UMAP Journal* 14(2): 113–135. Reprinted in *UMAP Modules: Tools for Teaching 1993,* edited by Paul J. Campbell, 171–208. Lexington, MA: COMAP, 1994.

Weissglass, Julian. 1993. Small-group learning. *American Mathematical Monthly* 100 (7): 662–668.

About the Authors

Steven Kolmes received his B.S. in zoology from Ohio University and his M.S. and Ph.D. degrees in zoology from the University of Wisconsin at Madison. He is currently the Rev. John Moltes, C.S.C, Chair in Science at the Univeristy of Portland. He is interested in behavioral ecology at the pest-pesticide interface and in efficiency theory in social insects.

D. Brooks McKinney received his B.S. in geology from Beloit College and his Ph.D. in petrology from Johns Hopkins University. He is interested in the crystallization of magmas and the geology of the Adirondack Mountains.

Kevin Mitchell received his B.A. in mathematics and philosophy from Bowdoin College and his Ph.D. in mathematics from Brown University. He is interested in hyperbolic tilings and algebraic geometry.

Hobart and William Smith's emphasis on and support for multidisciplinary work has permitted all three authors to explore new areas. Parts of this unit were first developed by McKinney for use in the Colleges' Department of Geoscience. Kolmes and Mitchell adapted the material for their team-taught integrated science and mathematics course for the Colleges' five-week intensive Summer Academic Orientation Program.

UMAP

Modules in Undergraduate Mathematics and Its Applications

Published in cooperation with the Society for Industrial and Applied Mathematics, the Mathematical Association of America, the National Council of Teachers of Mathematics, the American Mathematical Association of Two-Year Colleges, The Institute of Management Sciences, and the American Statistical Association.

Module 741

Waves and Strong Tides

L.R. King

Applications of Oceanography to Multivariate Calculus

COMAP, Inc., Suite 210, 57 Bedford Street, Lexington, MA 02173 (617) 862–7878

INTERMODULAR DESCRIPTION SHEET:	UMAP Unit 741
TITLE:	Waves and Strong Tides
AUTHOR:	L.R. King Dept. of Mathematics Davidson College Davidson, NC 28036 `riking@davidson.edu`
MATHEMATICAL FIELD:	Multivariate calculus
APPLICATION FIELD:	Oceanography
TARGET AUDIENCE:	Students in second- or third-semester calculus, differential equations, or calculus-based physics.
ABSTRACT:	We analyze the motion of water underneath a surface wave in the ocean in order to determine formulas for the wave speed. Next, we consider an idealized bay and use the same analysis to show that the length of the bay determines which waves can "fit" naturally in it. If any one of these waves has a period close to that of the tide, then this bay will resonate with the tide and experience a huge tidal range. Thus, changing the length of such a bay could cause dramatic changes in the tidal range. This result is used to illuminate the subject of harnessing the huge tides in the Bay of Fundy/Gulf of Maine system. An appendix on partial derivatives reviews the definition of a partial derivative and derives a particular equation, the continuity equation, that is used in the Module.
PREREQUISITES:	Differentiation of trigonometric (circular and hyperbolic) functions and partial differentiation of functions of two variables.

COMAP, Inc., Suite 210, 57 Bedford Street, Lexington, MA 02173
(800) 77–COMAP = (800) 772–6627 (617) 862–7878

Waves and Strong Tides

L.R. King
Dept. of Mathematics
Davidson College
Davidson, NC 28036
riking@davidson.edu

Table of Contents

Modules and Monographs in Undergraduate Mathematics and its Applications (UMAP) Project

The goal of UMAP is to develop, through a community of users and developers, a system of instructional modules in undergraduate mathematics and its applications, to be used to supplement existing courses and from which complete courses may eventually be built.

The Project was guided by a National Advisory Board of mathematicians, scientists, and educators. UMAP was funded by a grant from the National Science Foundation and is now supported by the Consortium for Mathematics and Its Applications (COMAP), Inc., a nonprofit corporation engaged in research and development in mathematics education.

Paul J. Campbell	Editor
Solomon Garfunkel	Executive Director, COMAP

1. Introduction

The prospect of an energy source that is clean and reliable and which could theoretically generate the same amount of electricity as 250 large nuclear power plants should attract attention. It has. This source is the tides in the Bay of Fundy in eastern Canada (see **Figure 1**).

Figure 1. Low and high tide in the Bay of Fundy.

Nowhere on Earth is the tide stronger. At its fastest, the incoming tide rises a foot every ten minutes. The tidal range exceeds 36 ft in some parts of the bay. Interest in harnessing this power is not new; early settlers built mills along the Bay that used tidal power.

Recent studies have changed the understanding of the cause of the tides in the Bay. Greenberg [1987] describes a mathematical model to test the effect of the various engineering projects proposed to convert tidal energy to large amounts of electricity. In what follows, we hope to give some understanding of how such a mathematical model might work.

2. Tides

Even today, tides elude a complete scientific explanation.

The sun and the moon "cause" the tides; but a host of local influences, such as ocean bottom topography, depth, coastline geometry, and the Earth's rotation, conspire to shape the appearance of the tide uniquely to any particular location. In our study, we limit ourselves to a rectangular bay of uniform depth and only consider the lunar tide, the tide caused by the moon.

The lunar tide is the result of the moon's gravitational effect on the oceans; it creates a pair of bulges (high tides) nearest the moon and on the side opposite it. If the moon did not move relative to the Earth, the lunar high tides would occur every 12 hrs. As it is, the lunar tidal period is about 12 hr 25 min.

A key to understanding why the tides are so powerful in the Bay of Fundy is the phenomenon of *resonance.* When considered as part of a larger system (the Bay of Fundy/Gulf of Maine system), the Bay "resonates" with the lunar tides; the Bay is (nearly) "tuned into" the 12 hr 25 min period as a child is "tuned into" the natural frequency of the swing set upon learning how to "pump."

The reader can observe an analogous effect of resonance in a coffee cup [Keeports 1988]. The cup/coffee system is our bay and is "tuned into" certain wave periods naturally: Observe the wave pattern in the coffee set up by a gentle thump of your finger. Now we set up an analogue of tidal forcing by sliding the cup across a table surface. With just the right cup (styrofoam) and just the right surface, the cup will hum "the right tune" as it slides, reinforcing the wave pattern until the waves begin to catapult coffee into the air, several inches above the cup.

The hum is the result of a periodic slide-stick action of the cup as it moves on the table surface. Resonance occurs when the period of this action approximates a period of the waves that occur naturally in the cup.

Tides are waves. The time required for an average such wave to pass a position is the *tidal period.* By contrast with tides, waves reaching the shore from a storm at sea have periods measured in minutes, and waves observed in a lake typically have periods measured in seconds.

We will develop a formula relating wave speed to ocean depth. This turns out to play a crucial role in explaining how a bay system could be "tuned into" certain wave periods.

3. Surface Ocean Waves

The building blocks for wave forms are functions of the form

$$\eta(x,t) = \eta_0 \cos(kx - \omega t),$$

each of which represents a wave with

- *wave number* k,
- *wavelength* $\lambda = 2\pi/k$ (crest-to-crest distance),
- *frequency* ω,
- *period* $P = 2\pi/\omega$, and
- *amplitude* η_0.

The wave travels in the direction of increasing x when $\omega > 0$ and in the opposite direction when $\omega < 0$.

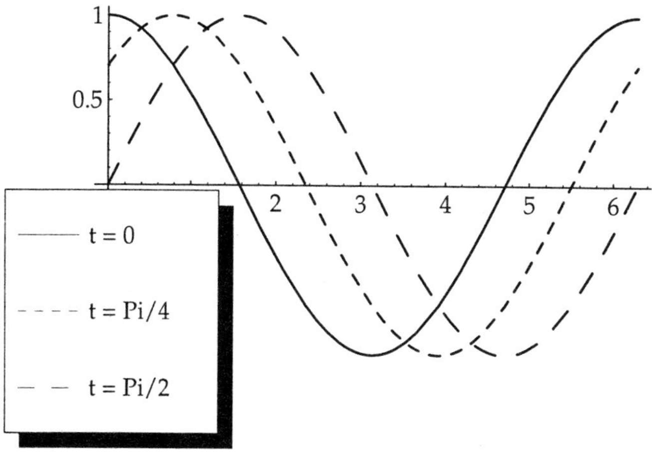

Figure 2. Progressive wave $z = \cos(x - t)$.

Most ocean waves do not have a sinusoidal profile like that in **Figure 2**. What we observe, however, is a combination of sinusoidal waves. To illustrate how this may be, **Figure 3** shows combinations of cosine graphs that approach the wave profile often used by cartoonists to depict the presence of water.

We construct the cartoonists' wave by piecing together end-to-end replicas of the graph of $y = 1 - \cos x$, for $-\pi/2 < x < \pi/2$, creating **Figure 3a**. The building blocks to approximate this curve are of the form $f_n(x) = a_n \cos 2nx$, for $n = 0, 1, 2, \ldots$. Successive approximations are

$$f_0, \qquad f_0 + f_1, \qquad f_0 + f_1 + f_2, \qquad \ldots .$$

The coefficients a_n are given by

$$a_0 = 1 - 2/\pi, \qquad a_n = \frac{4(-1)^n}{\pi(4n^2 - 1)}, \quad n = 1, 2, \ldots .$$

See **Figures 3b** and **3c**.

Two observations are in order here. The first is that the cartoonist is true to nature: A steady wind blowing over an otherwise calm ocean would produce waves with profile similar to the cartoonist's. Letting H denote the crest-to-trough distance and λ the wave length (crest-to-crest distance), the steepness H/λ most often is less that $1/10$, so that the actual wave profile is not nearly as steep as the cartoonist's. In fact, when H/λ reaches $1/7$, the wave breaks [Jelley 1989].

The second observation is that combining waves by adding, as we have done, is actually carried out in the ocean. That is, should two waves f and g encounter each other, the result is indeed $f + g$. This phenomenon is called the *principle of superposition*. **Figure 4** represents the superposition of ten waves, all with sinusoidal profile, with different wavelengths, directions, and periods. The waves were chosen more or less at random in an effort to simulate realistically the surface of an ocean.

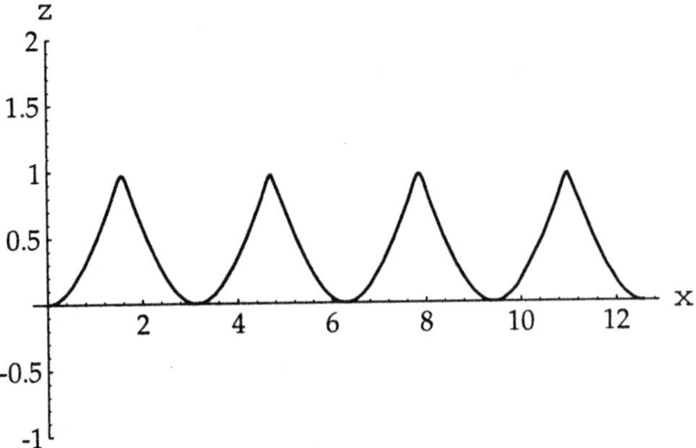

Figure 3a. Cartoonists' wave.

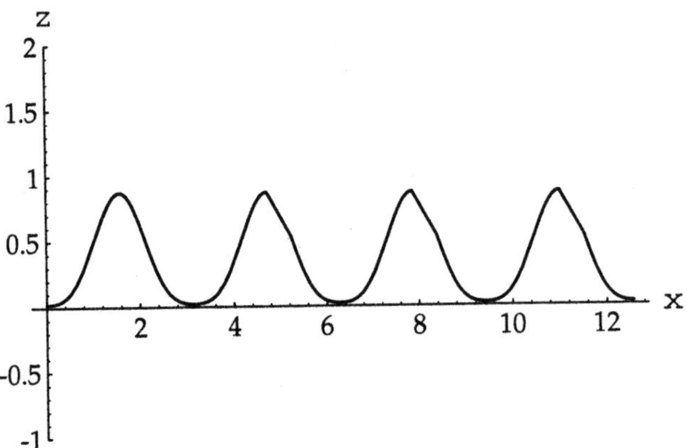

Figure 3b. Two-term approximation of cartoonists' wave.

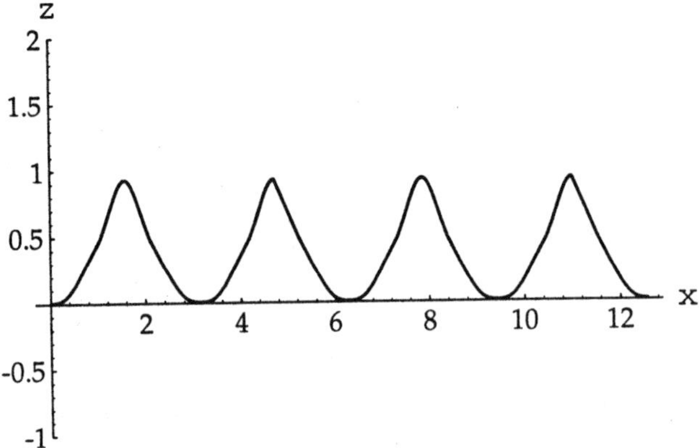

Figure 3c. Four-term approximation of cartoonists' wave.

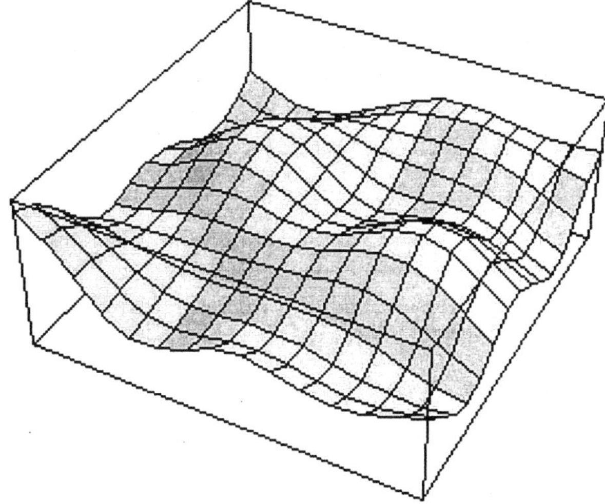

Figure 4. "Realistic" ocean by superposition of ten random sinsoidal waves.

An interesting case of superposition is when f and g have the same wave number k and the same frequency ω but travel in opposite directions. This happens when wave f is reflected off a wall and g is the reflection. The resulting combination $f+g$ is a *standing wave* (see **Figure 5**). It turns out that the tides may be considered as standing waves.

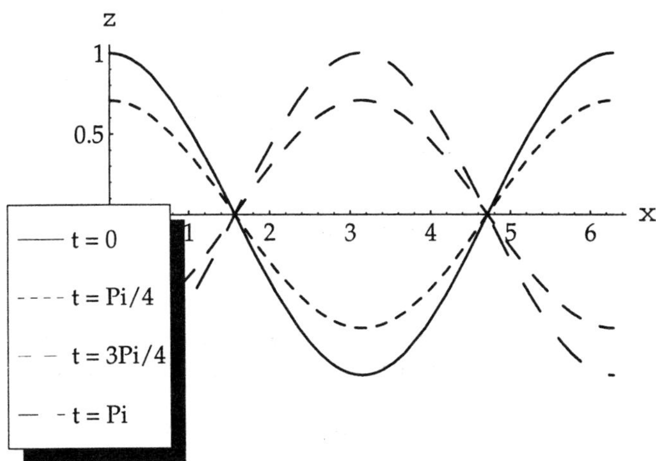

Figure 5. Standing wave $z = \cos x \cos t$.

The ocean puts a restriction on waves that distinguishes ocean waves from sound and light waves. A feature of this restriction is that deep-water waves travel at a speed that is dependent on the wavelength, unlike sound and light, so that longer waves travel faster than shorter ones. This fact is the basis for forecasts of storms: The long waves created by a storm reach the coast ahead of the other, shorter waves. These longest waves are detected by their purity (only one wavelength and therefore with a sinusoidal profile). These are the ground swells that tell of a storm at sea.

The restriction the ocean puts on $\eta(x,t) = \eta_0 \cos(kx - \omega t)$ is the

Dispersion relation $$c = \sqrt{gD}\sqrt{\frac{\tanh kD}{kD}},$$

where

$c = \omega/k$ is the wave speed,

$g = 9.8\ \mathrm{m/s^2}$ is the acceleration due to gravity, and

D is the ocean depth.

In the next sections, we will show how the ocean imposes this restriction.

Exercises

1. Show that the wave speed c equals λ/P, where P is the period.

2. Show that, as $x \to \infty$, we have $\tanh x \to 1$. In fact, show that if $D > \lambda$, then $|\tanh kD - 1| < 10^{-5}$. Recall that

$$\tanh(x) = \frac{e^x - e^{-x}}{e^x + e^{-x}}.$$

3. Use **Exercise 2** to show that when the wavelength λ is less than the ocean depth D, then the wave speed is approximated by a constant times the square root of the wavelength:

For deep-water waves: $$c \approx \sqrt{\frac{g}{2\pi}}\sqrt{\lambda}.$$

The relation of **Exercise 3** applies to *deep-water waves* and is the basis for the claim that waves created by a storm at sea disperse so that the longer waves travel faster.

What can be said of waves whose wavelength is large relative to the ocean depth, i.e., *shallow-water waves*?

Exercises

4. Show that, as $x \to 0$, we have $\tanh(x)/x \to 1$. In fact, show that if $\lambda > 400\pi D$, then $|\tanh(kD)/(kD) - 1| < 10^{-5}$.

5. Use **Exercise 4** to show that the wave speed of a shallow water wave (where λ is large relative to D) is approximated by a constant times the square root of the ocean depth:

$$\textbf{For shallow-water waves:} \qquad c \approx \sqrt{g}\sqrt{D}.$$

Waves created in the deep water that eventually travel to the shore must become shallow-water waves and slow down, so that c decreases (c is approximated by $\sqrt{gD}$, so c decreases when D does). Since such waves will keep the same period, the wavelength λ must decrease (see **Exercise 1**), forcing the steepness H/λ to increase. Bottom friction slows the trough more than the crest as the wave enters the sloping beach, so that the faster moving crest curls over to form the familiar breaker [Jelley 1989].

4. Surface Waves Feel the Ocean Floor

Let us assume that we have a surface wave in the form of a sinusoidal curve as in **Figure 6**, with crests perpendicular to the x-axis and travelling in the positive x-direction. Let us assume, further, that the ocean depth is constantly D. What restrictions are implied by the fact that the wave is an ocean wave? Can the wave "sense" the ocean floor? We will derive the dispersion relation of the pervious section to answer these questions.

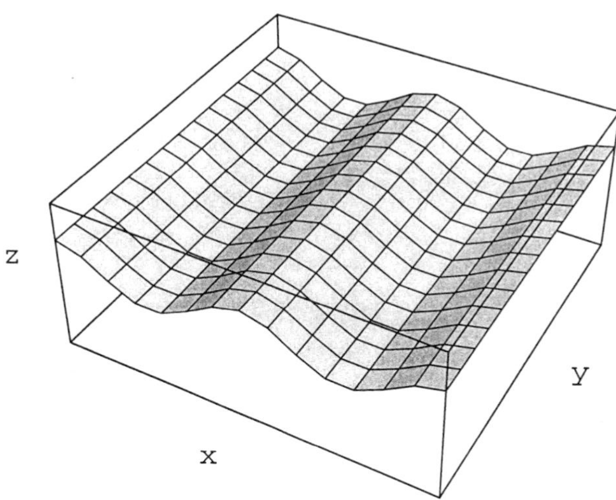

Figure 6. Sinusoidal ocean wave $\cos(x - t)$ for $t = 0$.

First, our assumption that the ocean can sustain such a wave form seems reasonable. Water at the crest will be tugged downward by gravity. Since it is not easily compressed, water in the neighboring trough must respond so as to make room. Because ocean water is not very viscous (sticky), this response is rather fast. Pressure and momentum allow the trough to become a crest and the process is repeated, making for a progressive wave.

To investigate how our wave senses the ocean floor, we consider the pressure beneath the wave, $p = p(x, z, t)$ (there is no dependence on y, since our wave form is the same for each fixed y). If we had a perfectly calm ocean, so that the sea surface is the plane $z = 0$, then the pressure depends only on the depth, $-z$, at (x, z) and would be given by $p = p(z) = -\rho g z$, where ρ is the (constant) density of seawater. In this situation, pressure and gravity are in balance, so that a parcel of water at any depth literally floats and is not accelerated in any direction. The result is perfectly still water from surface to ocean bottom.

The wave form amounts to a disturbance of the ocean surface. To highlight the dependency of the resulting pressure on this disturbance, we mark the departure from the balance between gravity and pressure given above for the undisturbed ocean and write

$$p = p(x, z, t) = -\rho g z + p^*(x, z, t).$$

The function p^* is called the *perturbation pressure.* We can say something about p^* immediately: Just at the ocean surface, the pressure should be atmospheric pressure (which for convenience we will consider to be 0).

Exercise

6. Show that at the surface of the ocean, the perturbation pressure is

$$p^*(x, \eta(x, t), t) = \rho g \eta(x, t).$$

The wave communicates with the deeper ocean via the perturbation function. Is this communication instantaneous? That is, does the deep ocean sense a higher p^* value when a crest of the wave is overhead, or is there a time delay? We make the assumption that there is no delay, that the ocean is rigid in this sense. Because of this and **Exercise 6** above, we assume that the perturbation function is of the form

$$p^* = p^*(x, z, t) = \rho g \eta(x, t) f(z),$$

for some function f that depends on the depth alone. Now we use some basic properties of ocean water to derive an equation that must be satisfied by p^* in order to determine f.

5. Equations for Motion Underneath a Surface Wave

Consider a water particle moving in our ocean so that at time t its position is $(x(t), z(t))$ (recall that in our ocean, motion is planar, in planes parallel to the x-z plane, $y = 0$). At each point (x, z) in our ocean and at each time t, the velocity of the water particle in the x-direction is $u = u(x, z, t) = dx/dt$ and in the z-direction is $w = w(x, z, t) = dz/dt$. See **Figure 7**.

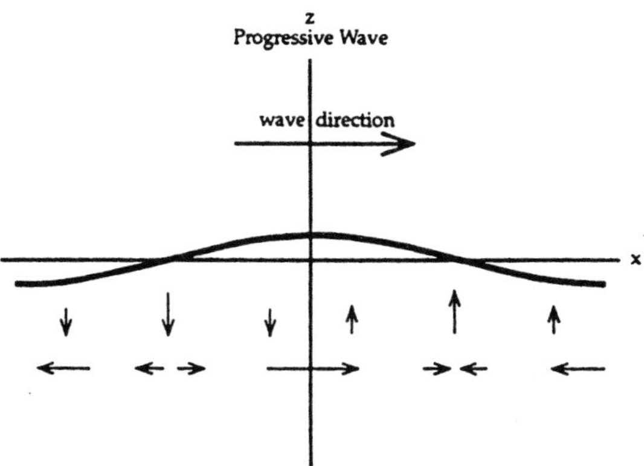

Figure 7. Horizontal arrows give u, vertical arrows give w.

Now let us follow a unit volume of water centered at (x, z) at time t. The momentum in the x-direction is ρu (unit mass times velocity) and its force is

$$\frac{\partial}{\partial t}(\rho u) = \rho\frac{\partial u}{\partial t} \qquad \text{(unit mass times acceleration).}$$

Note that we are assuming that the density of seawater is constant. The cause for this motion is the instantaneous difference in pressure in the x-direction created by our disturbance: $\partial p^*/\partial x$. Since water tends toward lower pressure, Newton's law $F = ma$ gives

First momentum equation: $$\rho\frac{\partial u}{\partial t} = -\frac{\partial p^*}{\partial x}.$$

Newton's law $F = ma$ for the z-direction is

Second momentum equation: $$\rho\frac{\partial w}{\partial t} = -\frac{\partial p^*}{\partial z}.$$

Taken together, these two equations are the *momentum equations* for our moving volume of water.

Now, sea water is not easily compressed. This means that the *net* flow rate of sea water away from a neighborhood of any point (x, z) in the ocean at any time t should be zero—any flow away must be accompanied by an equal flow into (see the **Appendix** for a derivation). Mathematically, this condition is expressed by the divergence of the flow being zero:

Continuity equation: $$\frac{\partial u}{\partial x} + \frac{\partial w}{\partial z} = 0.$$

6. Pressure Underneath a Surface Wave

In the exercises below, we derive our equation for p^* from the momentum equations and the continuity equation.

Exercises

7. Use the continuity equation to show that

$$\frac{\partial^2 u}{\partial t \partial x} = -\frac{\partial^2 w}{\partial t \partial z}.$$

8. Use the identities

$$\frac{\partial^2 u}{\partial x \partial t} = \frac{\partial^2 u}{\partial t \partial x}, \qquad \frac{\partial^2 w}{\partial z \partial t} = \frac{\partial^2 w}{\partial t \partial z}$$

and the momentum equations to show that

$$\frac{\partial^2 p^*}{\partial z^2} = -\frac{\partial^2 p^*}{\partial x^2}.$$

9. Use our assumption that $p * (x, z, t) = \rho g \eta_0 \cos(kx - \omega t) f(z)$ to show that

$$\frac{\partial^2 p^*}{\partial x^2} = -k^2 p^*.$$

Exercises 8 and **9** together give the equation that we seek:

$$\frac{\partial^2 p^*}{\partial z^2} - k^2 p^* = 0.$$

Because p^* is of the form $p^* = \rho g h(x, t) f(z)$, f must satisfy

$$\frac{\partial^2 f}{\partial z^2} - k^2 f = 0,$$

so that f must take the form

$$f(z) = Ae^{kz} + Be^{-kz}$$

for some constants A and B. We determine these constants from the boundary conditions.

At the ocean surface, we have

$$p^* = \rho g \eta(x, t) = \rho g \eta_0 \cos(kx - \omega t).$$

Since sea-surface elevations such as our wave are small compared to ocean depth, we take the surface condition to be when $z = 0$. This means that $f(0) = 1$.

At the ocean floor, there is no vertical flow, so that $w = w(x, -D, t) = 0$, for all x and all t.

Exercises

10. Use the second momentum equation to show that the ocean floor condition above implies that $f'(-D) = 0$.

11. Recall that $\cosh x = (e^x + e^{-x})/2$. Show that the boundary conditions at the sea surface and at the floor imply that

$$f(z) = \frac{\cosh k(z + D)}{\cosh kD}.$$

We have discovered that a surface wave does feel the ocean floor and that it does so via the perturbation pressure function,

$$p^*(x, z, t) = \rho g \eta_0 \cos(kx - \omega t) \frac{\cosh k(z + D)}{\cosh kD}.$$

This solution leads to the dispersion relation via some approximations at the sea surface. As before, we approximate the sea surface $\eta(x, t)$ by $z = 0$. The vertical flow $w(x, 0, t)$ is approximately

$$\frac{\partial}{\partial t}\eta(x, t),$$

as **Figure 8** shows.

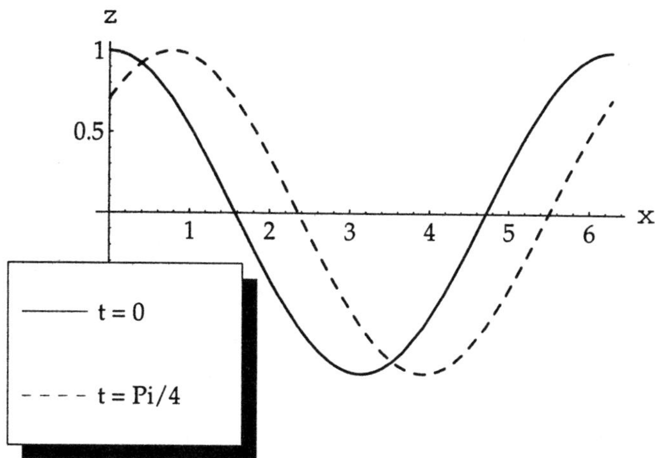

Figure 8. Vertical motion, w, at the sea surface.

Exercises

12. Use the approximation

$$w(x,0,t) = \frac{\partial}{\partial t}\eta(x,t)$$

to show that, when $z = 0$,

$$\rho\frac{\partial w}{\partial t} = -\rho\omega^2\eta(x,t).$$

13. Use the second momentum equation at $z = 0$ and **Exercise 12** to show that $\omega^2 = gk \tanh kD$.

14. Use **Exercise 13** to obtain the dispersion relation

$$c = \sqrt{gD}\sqrt{\frac{\tanh kD}{kD}}.$$

7. Natural Frequencies of a Narrow Bay

As noted in the introduction, the reason for the gigantic tide range at in the Bay of Fundy is the concept of resonance as it applies to the Bay of Fundy / Gulf of Maine bay system. Greenberg [1987] has concluded that the natural frequency of waves in this system is close to the frequency of the tides, so that the bay system resonates with the tides, creating the unusual tide range. The public issue of the desirability of harnessing this tidal power by introducing a dam at the head of the Bay of Fundy heightens the importance of the conclusions of the study.

The fact is that a natural frequency will be altered as the geometry of the bay system is altered. A dam, then, could bring the bay system even closer to perfect resonance with the tides with potentially disastrous consequences to distant points such as Boston. Even slight changes in the tidal response could alter entire ecosystems. For example, those areas that are alternately exposed and flooded by the tides could be changed so that there is no exposure at low tide or, on the other hand, so that the tide never reaches the area.

In what follows, we consider a very simple narrow bay system: rectangular, with one end open to the sea, with uniform depth D and uniform width W. Our purpose is to determine the natural frequency of waves in the system. We apply the equations developed thus far.

Let us assume that we have a bay system as in **Figure 9**.

Our bay is like a tuning fork. An initial disturbance from the state of equilibrium (calm) causes oscillations (waves) with certain frequencies determined by the geometry of the system. The initial disturbance, say caused by low barometric pressure, could be too complicated to be described by a single equation. Fortunately, there is a theorem in mathematics [Kaplan 1973, 482] that assures us that the disturbance can be approximated with arbitrarily good accuracy

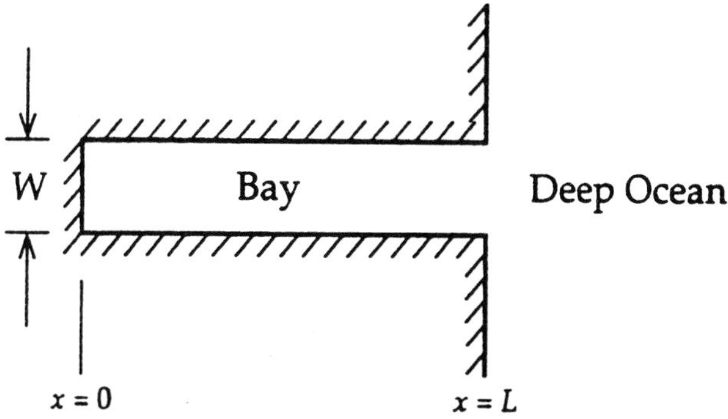

Figure 9. Uniform bay of depth $D = 20$ meters.

using sinusoidal waves and the principal of superposition. The approximation of the cartoonist's wave (see **Figure 2**) illustrates this theorem.

Our approach is to assume that the initial disturbance and the response are sinusoidal. Because of the principle of superposition and the theorem mentioned above, we lose no generality. Therefore, let us consider a wave such as we have considered above, with profile $\eta(x,t) = \eta_0 \cos(kx - \omega t)$, whose crests are perpendicular to the axis of the bay.

In order for this wave to be aware of the existence of the bay, it must meet certain conditions:

- First, because the bay is shallow in comparison to the wavelength, the wave speed c must be approximately $\sqrt{gD}$ (see **Exercise 5**).
- There must be no horizontal flow at the *head*, $x = 0$, of the bay; so that $u(0,z,t) = 0$ for all z and all t.
- Finally, water conditions along the line defining the *mouth* of the bay, $x = L$, should be the same as figured from the bay side, $x \leq L$, as from the ocean side, $x \geq L$.

The natural frequencies for the bay are derived from these three conditions.

The shallow water condition shows that the perturbation pressure is independent of depth! Because kD is small, the we have

$$\frac{\cosh k(z+D)}{\cosh kD} \approx 1,$$

so that $p^*(x,z,t)$ becomes

$$p^*(x,t) = \rho g \eta_0 \cos(kx - \omega t).$$

Exercises

15. Use the shallow-water approximation for wave speed $c \approx \sqrt{gD}$ for a bay of depth $D = 20$ m and length $L = 150$ km and a wave of period $P = 12$ hr 25 min to calculate how long such a wave would take to traverse the bay from mouth to head. This period P is the tidal period, so we are computing the time difference between high tide at the mouth of the bay and at the head.

16. With the same wave and bay of **Exercise 15** and **Figure 7**, show that the quantity $\cosh(k(z+D))/\cosh(kD)$ differs from 1 by no more that 3×10^{-8}.

We now have an apparent contradiction: By assuming the perturbation pressure is independent of z, because of the second momentum equation

$$\rho \frac{\partial w}{\partial t} = -\frac{\partial p^*}{\partial z},$$

we have $\partial w/\partial t = 0$, indicating that the vertical component of the motion of the sea water is constant over time! How could this be? The tide *rises and falls*, after all! Consider a tidal range (from low to high tide) of 2 m; this vertical motion occurs over a period of about 6 hr 12 min, so the average rate of vertical motion during this period is 2 m per 6 hr 12 min, or 2 m per 22,320 sec. Thus, the average value of w over this period is 2/22,320 m/sec, or less that 10^{-4} m/s. The time rate of change, $\partial w/\partial t$, is even smaller, and is ignored in the computations to determine the natural frequency of the bay.

We learn in **Exercise 17** below that no one of our wave forms $\eta(x,t) = \eta_0 \cos(kx - \omega t)$ can "fit" in our bay, because such a wave cannot meet the condition of no horizontal flow at the head. In this instance, we *must* consider sums of such wave forms and use the principle of superposition.

It turns out that the simplest approach works: The superposition of waves with cosine terms $\cos(kx - \omega t)$ and $\cos(kx + \omega t)$ produces a standing wave that can "fit" and meet the no-flow condition. That this should be the case is not surprising: The incoming wave $\eta_0 \cos(kx + \omega t)$ is reflected at the head of the bay, creating a standing wave by combining itself with its reflected wave: $\eta_0 \cos(kx + \omega t) + \eta_0 \cos(kx - \omega t)$.

Exercise

17. Use the first momentum equation with our shallow-water perturbation pressure $p^*(x,t) = \rho g \eta_0 \cos(kx - \omega t)$ to show that at $x = 0$,

$$\frac{\partial u}{\partial t} = kg\eta_0 \sin(kx - \omega t).$$

Since this term is not 0 for all t, then $u(0,t)$ cannot be 0 for all t.

We show in **Exercises 18** and **19** below that standing waves of the form

$$\eta(x,t) = \eta_0 \cos kx \cos \omega t$$

do satisfy the condition of no horizontal flow at the head $x = 0$ of the bay, and we derive the equation for the horizontal flow function $u(x,t)$. Note that just as the perturbation pressure is independent of depth in our bay, so is the horizontal flow.

Exercises

18. Use the first momentum equation and $p^*(x,t) = \rho g \eta(x,t)$ (where η is the standing wave above) to show that

$$\frac{\partial u}{\partial t} = kg\eta_0 \sin kx \cos \omega t.$$

Assume that initially there was no flow in the bay, $u(x,0) = 0$ for all x, to show that

$$u(x,t) = \frac{c}{D}\eta_0 \sin kx \sin \omega t.$$

19. Use **Exercise 18** to show that the condition of no horizontal flow at the head of the bay is satisfied.

It is interesting that our choice of coordinates (head of the bay is $x = 0$ and mouth of the bay is $x = L$) is such that the condition at the head of the bay puts no restriction on k. The condition to be satisfied at the mouth puts the restriction on k that, in effect, tells us the natural frequencies of the bay.

Consider the imaginary line defining the mouth of the bay. The perturbation pressure here, from the point of the bay, is $p^* = \rho g \eta$. As remarked above, this should be the same from the point of view of the open ocean. From the point of view of the bay, the rate of movement of water across this line is $\rho A u$, where $A = WD$, the cross section of the bay, and u is the flow rate. Again, this should be the same from the point of view of the open sea. To rid ourselves of the amplitude of the wave, we consider the ratio, $p^*/\rho A u$, of these two quantities. Let Z_{bay} be this ratio from the bay point of view and Z_{ocean} be the comparable entity from the ocean point of view.

Exercise

20. Show that $Z_{\text{bay}} = \frac{\sqrt{g}}{W\sqrt{D}} \cot kL \cot \omega t$.

Now, Z_{ocean} should be zero, as the width W and the depth D are large from the ocean side. The equality $Z_{\text{bay}} = Z_{\text{ocean}}$ gives the condition that we seek to find the natural frequencies of our bay:

$$kL = \left(n + \frac{1}{2}\right)\pi, \qquad n = 0, 1, 2, \ldots .$$

This restriction on the wave number k becomes a restriction on the frequency of our standing wave via the shallow-water approximation for the wave speed, $c = \omega/k \approx \sqrt{gD}$, so that

$$\omega L = \left(n + \frac{1}{2}\right)\pi c, \qquad n = 0, 1, 2, \ldots .$$

Thus, the natural frequencies for a uniform bay of length L are:

$$\omega = \frac{\left(n + \frac{1}{2}\right)\pi c}{L}, \qquad n = 0, 1, 2, \ldots .$$

Exercises

21. Show that the longest wavelength associated with a wave that "fits" our uniform bay is four times the length of the bay.

22. Show that a uniform bay of depth 20 m that is 156.45 km long resonates "perfectly" with the tide. That is, if $L = 156,450$ m, the natural frequency given by the equation above with $n = 0$ of the bay is $2\pi/P$, where P is the tidal period 12 hr 25 min.

The result of **Exercise 22** indicates that a uniform bay about 150 km long will have sizable tides—the natural "sloshing" frequency is closely matched by the gentle periodic pushing by the tide at the mouth of the bay. Our ideal uniform bay provides some insight on how the plans to harness tidal power in the Bay of Fundy by building a large dam might cause a change in the tides in the Bay. Such a dam in our uniform bay would alter its length and therefore move the natural frequencies of the shortened bay either nearer or farther away from the tidal frequency and thereby change the response either nearer resonance or farther away.

8. Solutions to the Exercises

1.

$$c = \omega/k = \frac{2\pi/P}{2\pi/\lambda} = \lambda/P.$$

2. The first result follows from $\lim_{x\to\infty} e^{-x} = 0$. If $D > \lambda$, then $kD = 2\pi(D/\lambda) > 2\pi$, and $\tanh(2\pi) \approx 0.999993025$.

3.

$$c = \sqrt{gD}\sqrt{\frac{\tanh kD}{kD}} \approx \frac{g}{k} = \sqrt{\frac{g}{2\pi/\lambda}} = \sqrt{\frac{g}{2\pi}}\sqrt{\lambda}.$$

4. For the first part, apply l'Hôpital's Rule. If $\lambda > 400\pi D$, then $kD = (2\pi/\lambda)D < 1/200$; observe that $\tanh(1/200)/(1/200) \approx 0.9999916668$.

5.

$$c = \sqrt{gD}\sqrt{\frac{\tanh kD}{kD}} \approx \sqrt{gD} \qquad \text{when} \qquad \frac{\tanh kD}{kD} \approx 1;$$

$kD = 2\pi D/\lambda$ is small when λ is large relative to D.

6. $0 = p(x, \eta(x,t), t) = -\rho g \eta(x,t) + p * (x, \eta(x,t), t)$.

7. Differentiate both sides of the continuity equation partially with respect to t.

8. From the two momentum equations, we get respectively

$$\rho\frac{\partial^2 u}{\partial x \partial t} = -\frac{\partial^2 p^*}{\partial x^2} \qquad \text{and} \qquad \rho\frac{\partial^2 w}{\partial z \partial t} = -\frac{\partial^2 p^*}{\partial z^2}.$$

The result follows from **Exercise 7** and the identities given.

9. Just a computation.

10. Since $w(x, -D, t) = 0$ for all t, the second momentum equation gives

$$\partial w/\partial t = -\eta_0 g \cos(kx - \omega t) f'(-D) = 0$$

for all x and all t, so $f'(-D)$ must be 0.

11. $f(z) = Ae^{kz} + Be^{-kz}$ and $f(0) = 1$ imply that $A + B = 1$; $f'(-D) = 0$ implies that $A = e^{kD}/\left(e^{kD} + e^{-kD}\right)$ and $1 - A = e^{-kD}/\left(e^{kD} + e^{-kD}\right)$; and substitution gives the result.

12.

$$\frac{\partial}{\partial t}\eta(x,t) = \eta_0 \omega \sin(kx - \omega t), \qquad \text{so, when } z = 0,$$

$$\rho\frac{\partial w}{\partial t} = \rho(-\eta_0 \omega^2 \cos(kx - \omega t)) = -\rho\omega^2 \eta(x,t).$$

13. When $z = 0$, the second momentum equation and **Exercise 12** give

$$-\rho\omega^2 \eta_0 \cos(kx - \omega t) = -\rho g k \eta_0 \cos(kx - \omega t)\,\frac{\sinh kD}{\cosh kD}.$$

14.

$$c^2 = \frac{\omega^2}{k^2} = \frac{gk \tanh kD}{k^2} = \frac{gD \tanh kD}{kD}.$$

15. $c \approx \sqrt{9.8}\sqrt{20} = 14$ m/sec; 150,000/14 sec $\approx$ 2 hr 59 min.

16.

$$\left|\frac{\cosh k(z+D)}{\cosh kD} - 1\right| \le \left|\frac{1 - \cosh kD}{\cosh kD}\right| \approx 2.06 \times 10^{-8},$$

since, using the fact that $\omega/k = c = 14$, we have

$$kD = (\omega/14)D = \frac{2\pi/44,700}{14}\cdot 20 = \frac{2\pi}{31,290}.$$

17. Compute $(-1/\rho)\partial p^*/\partial x$.

18. For the first part, compute $(-1/\rho)\partial p^*/\partial x$;

$$u(x,t) = u(x,t) - u(x,0) = \int_0^1 \frac{\partial u}{\partial t}(x,s)\,ds; \qquad (k/\omega)g = g/c = c/D.$$

19. Substitute $x = 0$ into **Exercise 18**.

20. Note that $g/Wc = \sqrt{g}/W\sqrt{D}$.

21. Solve for λ: $(2\pi/\lambda)L = \pi/2$.

22. $$\frac{2\pi}{\omega} = \frac{2\pi L}{\frac{1}{2}\pi c} = \frac{(4)(156,450)}{\sqrt{9.8}\sqrt{20}} = 44,700 \text{ sec } = 12 \text{ hr } 25 \text{ min.}$$

23. $$\frac{(\Delta u + \Delta v + \Delta w)(t-t_0)A}{(x-x_0)(y-y_0)(z-z_0)(t-t_0)} = \frac{\Delta u}{x-x_0} + \frac{\Delta v}{y-y_0} + \frac{\Delta w}{z-z_0}.$$

Now use the definition of the partial derivative.

9. Appendix on Partial Derivatives

Consider the motion of water created by a surface wave. At any given time t and at any point (x, y, z) in the ocean beneath the wave, the velocity of the water is given by its components in the x-, y-, and z-directions u, v, and w, respectively. We write $u = u(x,y,z,t)$, $v = v(x,y,z,t)$, and $w = w(x,y,z,t)$. Now fix a time $t = t_0$ and a point (x_0, y_0, z_0). The time *partial* derivative of u is the derivative of the function f obtained from u by holding the space variables x, y, and z constant:

$$\frac{\partial u}{\partial t} = f'(t_0),$$

where $f(t) = u(x_0, y_0, z_0, t)$. Thus,

$$\frac{\partial u}{\partial t} = \lim_{t \to t_0} \frac{f(t) - t(t_0)}{t - t_0} = \lim_{t \to t_0} \frac{u(x_0, y_0, z_0, t) - u(x_0, y_0, z_0, t_0)}{t - t_0}.$$

From this, it follows that over a small time interval from t_0 to t, the change in the x-component of the water velocity $u(x_0, y_0, z_0, t) - u(x_0, y_0, z_0, t_0)$ is approximately $\partial u/\partial t(t - t_0)$.

The spatial *partial* derivative of u in the x-direction is the derivative of the function g obtained from u by holding y, z, and t constant:

$$\frac{\partial u}{\partial x} = g'(x_0),$$

where $g(x) = u(x, y_0, z_0, t_0)$. Thus,

$$\frac{\partial u}{\partial x} = \lim_{x \to x_0} \frac{g(x) - g(x_0)}{x - x_0} = \lim_{x \to x_0} \frac{u(x, y_0, z_0, t_0) - u(x_0, y_0, z_0, t_0)}{x - x_0}.$$

From this, it follows that over a small space interval in the x-direction from x_0 to x, the change in the x-component of the water velocity $u(x, y_0, z_0, t_0) - u(x_0, y_0, z_0, t_0)$ is approximately $\partial u/\partial x(x - x_0)$.

The other partial derivatives of u, $\partial u/\partial y$ and $\partial u/\partial z$, are defined in like manner, as are the partial derivatives of the other component functions of the water velocity.

We can now use the approximations cited above to derive the *continuity equation* for water motion underneath a surface wave (our use of this equation assumed that there was no water motion in the y-direction, so that $\partial v/\partial y = 0$). To show that this equation holds at a particular point (x_0, y_0, z_0) at a particular time $t = t_0$, we imagine a tiny cube whose opposite corners are (x_0, y_0, z_0) and (x, y, z), as in **Figure 10**. The idea is to exploit the fact that sea water is not easily compressed, so that we may assume that it is of constant density; this means that mass and volume are proportional.

Now, over any time interval, the *mass* of water entering our tiny cube must equal the mass that leaves it. Because of our assumption of incompressibility, we may claim that over any time interval that the *volume* of water entering our cube must equal the volume leaving.

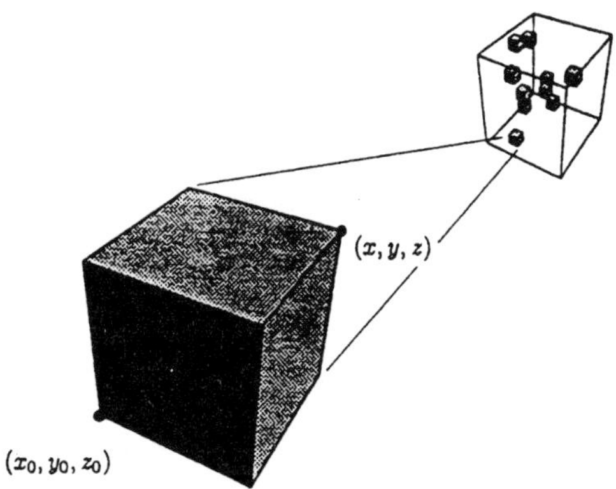

Figure 10. Cube of water (enlarged) at depth.

The volume leaving the right face of our cube during the time interval from t_0 to t, assuming the interval is small, is approximately $u(x, y_0, z_0, t_0)(t - t_0)A$, where A is the area of the face (this could be negative if the x-component of the water motion u were negative, in which case the absolute value would be the volume of the water entering). The volume entering the left face of our cube in this same time interval is approximately $u(x_0, y_0, z_0, t_0)(t - t_0)A$. The difference,

$$(u(x, y_0, z_0, t_0) - u(x_0, y_0, z_0, t_0))(t - t_0)A,$$

must be compensated for by the corresponding differences involving the other two pairs of faces, so that the following sum S is approximately 0:

$$(\Delta u + \Delta v + \Delta w)(t - t_0)A,$$

where

$$\begin{aligned}
\Delta u &= u(x, y_0, z_0, t_0) - u(x_0, y_0, z_0, t_0), \\
\Delta v &= v(x_0, y, z_0, t_0) - v(x_0, y_0, z_0, t_0), \\
\Delta w &= w(x_0, y_0, z, t_0) - w(x_0, y_0, z_0, t_0).
\end{aligned}$$

Exercise

23. Show that dividing S by $(t - t_0)((x - x_0)(y - y_0)(z - z_0)$ and shrinking our tiny cube to its corner point (x_0, y_0, z_0) by allowing for $x \to x_0$, $y \to y_0$, and $z \to z_0$, we get the continuity equation.

References

Bascom, Willard. 1964. *Waves and Beaches.* Science Study Series s34. Garden City, NY: Doubleday.

Clancy, Edward P. 1969. *The Tides.* Science Study Series s56. Garden City, NY: Doubleday.

Garrett, Chris, and Leo R.M. Maas. 1993. Tides and their effects. *Oceanus* 36 (Spring 1993): 27–37.

Gill, Adrian E. 1982. *Atmosphere-Ocean Dynamics.* Orlando, FL: Academic.

Greenberg, David A. 1987. Modeling tidal power. *Scientific American* 257 (5) (November 1987): 128–131.

Jelley, J.V. 1986. The tides, their origins and behavior. *Endeavour* 10: 184–190.

________. 1989. Sea waves: Their nature, behaviour,m and practical importance. *Endeavour* 13: 148–156.

Kaplan, Wilfred. 1973. *Advanced Calculus.* 2nd ed. Reading, MA: Addison-Wesley.

Keeports, David. 1988. Standing waves in a styrofoam cup. *Physics Teacher* 26 (October 1988): 456–457.

Sverdrup, H.U., et al. 1942. *The Oceans: Their Physics, Chemistry, and General Biology.* New York: Prentice-Hall.

Thurston, Harry. 1985. Fundy's fecund barrens. *Audubon* 87 (5) (September 1985): 88–103.

Vorobyov, Ivan. 1994. The bounding main. *Quantum* 4 (May-June 1994): 20–25.

About the Author

L.R. King received a B.S. from Davidson College and a Ph.D. from Duke University (dimension theory). His interest in advertising oceanography in undergraduate mathematics courses is the result of spending a sabbatical year in the School of Oceanography of the University of Washington in 1988–89.

UMAP

Modules in Undergraduate Mathematics and Its Applications

Published in cooperation with the Society for Industrial and Applied Mathematics, the Mathematical Association of America, the National Council of Teachers of Mathematics, the American Mathematical Association of Two-Year Colleges, The Institute of Management Sciences, and the American Statistical Association.

Module 744

Graphs, Digraphs, and the Rigidity of Grids

Brigitte Servatius

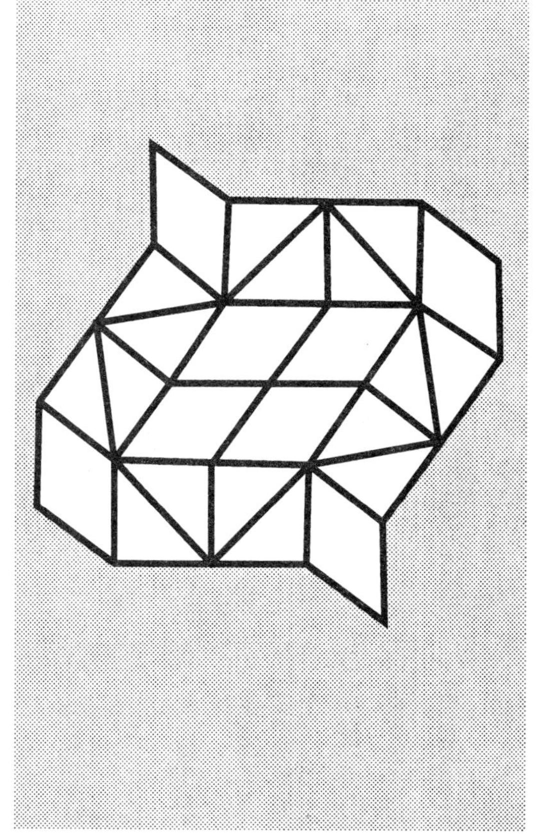

Applications of Graph Theory and Linear Algebra to Architecture and Statics

COMAP, Inc., Suite 210, 57 Bedford Street, Lexington, MA 02173 (617) 862–7878

INTERMODULAR DESCRIPTION SHEET: UMAP Unit 744

TITLE: Graphs, Digraphs, and the Rigidity of Grids

AUTHOR: Brigitte Servatius
Mathematical Sciences Dept.
Worcester Polytechnic Institute
Worcester, MA 01609
`bservat@wpi.wpi.edu`

MATHEMATICAL FIELD: Graph theory, linear algebra

APPLICATION FIELD: Architecture, statics

TARGET AUDIENCE: Students in discrete mathematics.

ABSTRACT: We study the rigidity of frameworks made of rigid rods in the plane that are connected by two-dimensional ball joints; in particular, we consider grids of squares. We discuss strategies for bracing grids with rods, and we formulate criteria for rigidity of the grid in terms of linear algebra and in terms of the brace graph of the grid. We briefly consider efficiency and fault tolerance of bracings. Bracing a grid with cables rather than rods leads to a tensegrity structure and to criteria for rigidness in terms of a system of inequalities and in terms of the cable-brace directed graph of the grid. An appendix gives suggestions for the instructor.

PREREQUISITES: Solution of simple simultaneous linear equations.

COMAP, Inc., Suite 210, 57 Bedford Street, Lexington, MA 02173
(800) 77–COMAP = (800) 772–6627 (617) 862–7878

Graphs, Digraphs, and the Rigidity of Grids

Brigitte Servatius
Mathematical Sciences Dept.
Worcester Polytechnic Institute
Worcester, MA 01609
bservat@wpi.wpi.edu

Table of Contents

Modules and Monographs in Undergraduate Mathematics and its Applications (UMAP) Project

The goal of UMAP is to develop, through a community of users and developers, a system of instructional modules in undergraduate mathematics and its applications, to be used to supplement existing courses and from which complete courses may eventually be built.

The Project was guided by a National Advisory Board of mathematicians, scientists, and educators. UMAP was funded by a grant from the National Science Foundation and is now supported by the Consortium for Mathematics and Its Applications (COMAP), Inc., a nonprofit corporation engaged in research and development in mathematics education.

Paul J. Campbell	Editor
Solomon Garfunkel	Executive Director, COMAP

1. Rigidity of Frameworks

1.1 Frameworks

There is nothing like news of an earthquake to make us wonder about the strength and rigidity of buildings and bridges. For answers, we must depend upon the skill of architects and civil engineers; and they, in turn, depend upon experience and mathematics.

Consider two simple structures made of solid rods joined at their endpoints, one structure in the form of a triangle and the other a square. For the triangle, the three rod lengths determine the three angles; so the only ways that a triangular structure can fail are to break a bond at one corner or to break a rod (see **Figure 1**). A triangle is *rigid*.

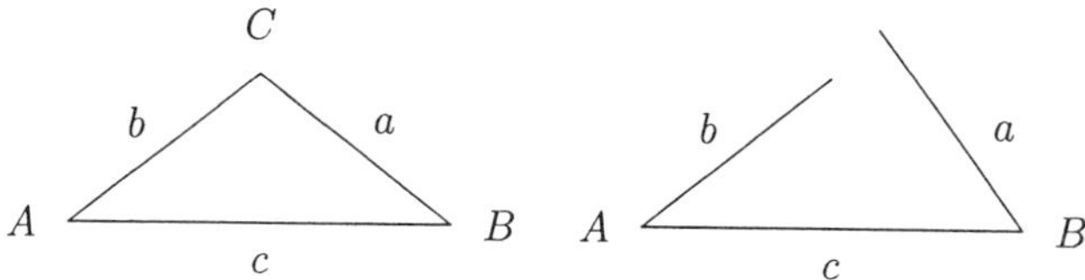

Figure 1. A triangle, which is rigid, and a triangle with a bond broken at one corner.

A square, however, may be *deformed* into a rhombus without breaking any bonds or bending any rods (see **Figure 2**).

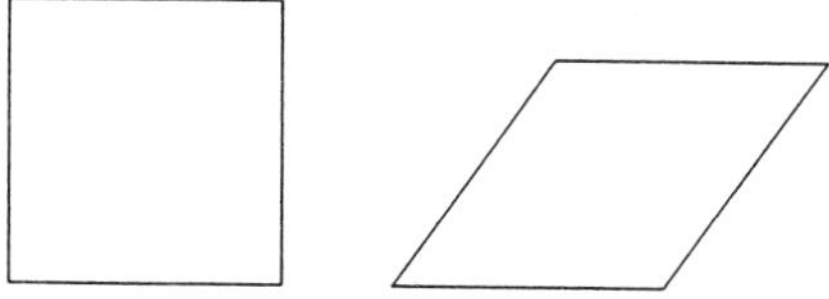

Figure 2. A square, which is not rigid, can be deformed into a rhombus.

Some bonds, for instance, from gluing or welding, will fix angles to some extent; but this should not be relied upon. Even if the rivets and the welds are very strong, they will do a much better job of keeping the objects pinned together than of fixing the angles. For instance, if you nail the ends of two ordinary boards together, then it should take several hundred pounds of force to break the bond (on the order of 300 lbs/nail), usually by snapping the nails or splitting the wood. On the other hand, it is not nearly so hard to bend and twist the joint, since that can be done by only bending and twisting the nails. (One reason that this is so easy is that each of the two boards acts as a lever with respect to the joint.)

Since a simple joint does a relatively poor job of fixing angles, it is safer to assume simply that the joints prevent the bars from separating but are completely flexible with regard to twisting and turning. Such a joint is called a *ball joint*. A familiar example is the joint in your shoulder. You also have a ball joint in your knee; this joint, however is only two-dimensional and resists twisting from the side, the source of numerous sports injuries.

In this Module, we study the rigidity of frameworks made of rigid rods in the plane that are connected by two-dimensional ball joints.

1.2 Walls and Grids

A modern wall often consists of a thin, light surface material covering a combination of insulations and backed by a rigid framework. Non-wasteful use of lumber or steel requires that the rods in the framework not be too long. A natural choice for a framework is a grid of squares, as pictured in **Figure 3**.

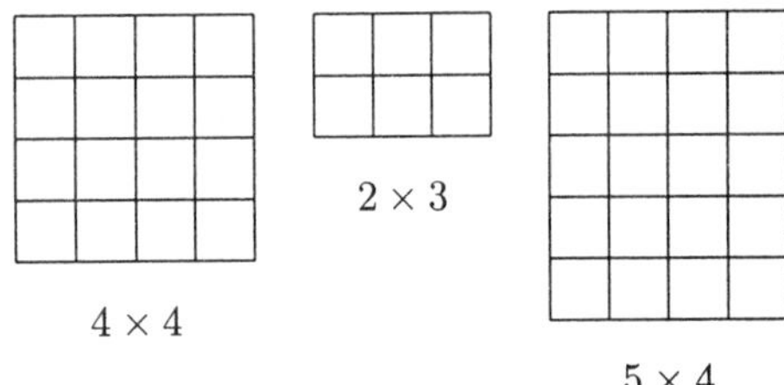

Figure 3. Grids of squares.

Each square in the grid is made up of four short rods joined at the four corners. Let us call the vertical rods *posts*, and the horizontal rods *beams*.

We want the grid framework to be rigid in the plane of the wall. Rigidity in a plane is easier to achieve than rigidity in three-space; many frameworks rigid in the plane will fold up in three-space. A rectangular grid of squares, however, is not rigid even in the plane (see **Figure 4**).

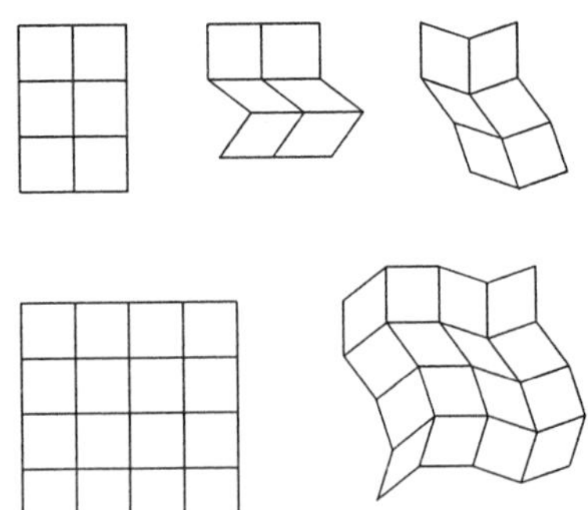

Figure 4. Grids in the plane and deformations of them.

It is important to notice that in any deformation, all the posts in the same row are parallel, since they are the opposite sides of rhombi. Similarly, all the beams in each column must be parallel in any deformation.

Fortunately, we can rigidify the grid by adding diagonal braces at each square. The fully braced grid is rigid in the plane because it is made up of triangles touching along edges, and triangles *are* rigid (see **Figure 5**).

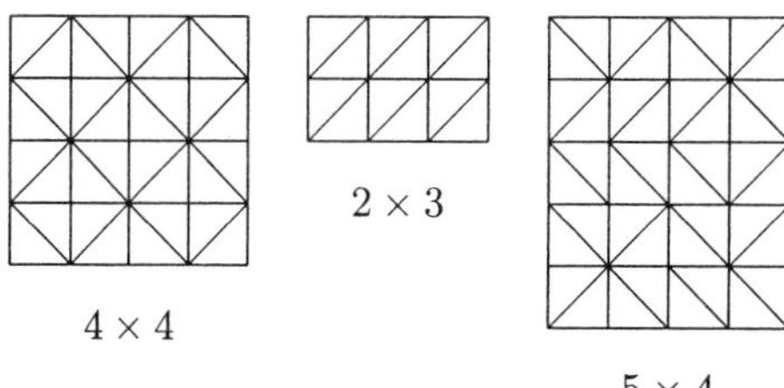

Figure 5. Fully braced grids.

1.3 Bracing a Grid

When considering a deformation of a grid, in order to specify our frame of reference, we will always fix the top-left post. In particular, this means that all the posts in the top row will be vertical.

We want to brace a grid so that it is rigid in the plane. If we brace every square, then the grid is clearly rigid; however, **Figure 6** shows that sometimes we can do the job with fewer braces.

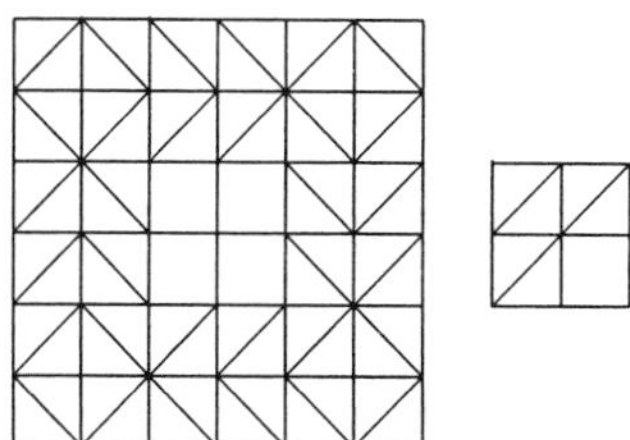

Figure 6. Rigid braced grids.

In the grids in **Figure 6**, every unbraced square has a corner surrounded by braced squares, so the angle at the corner must be a right angle, and so that square cannot be deformed. The main question is:

Which choices of braces rigidify a grid?

Let ρ_i be the angle that the posts in row i make with the horizontal, and and κ_j be the angle that the beams in column j make with the horizontal. Then $\rho_i - \kappa_j$ is the angle of the bottom-left corner of the rhombus (i, j) (see **Figure 7**).

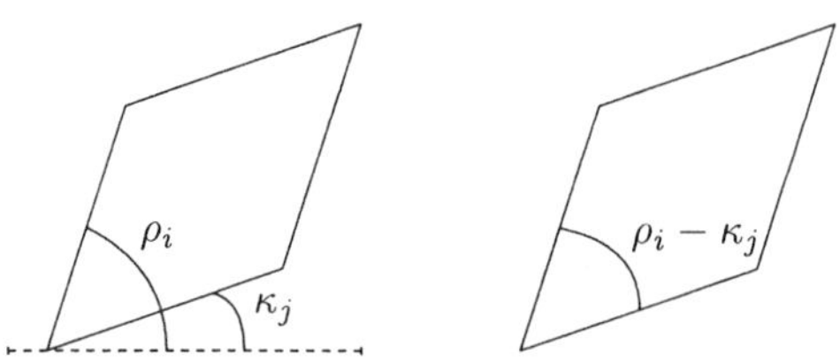

Figure 7. Row and column angles.

To compare the sizes of these angles, let's assume that we are looking for small deformations, say with $-20^\circ < \kappa_j < 20^\circ$ and $70^\circ < \rho_i < 110^\circ$. Since a brace in square (i, j) forces the posts in row i to be perpendicular to the beams in column j, we have the equation

$$\rho_i = \kappa_j + 90^\circ.$$

Thus, every braced grid corresponds to a system of linear equations each like this one, together with the equation

$$\rho_1 = 90^\circ,$$

which comes from our choice of frame of reference.

We know that this linear system has at least one solution, namely $\rho_i = 90^\circ$ for every i, and $\kappa_j = 0^\circ$ for every j. If this is the only solution, then we may conclude that the grid is rigid, and vice versa. This means that rigidity may be decided simply by ordinary linear algebra.

Example 1. We can determine that the grid of **Figure 8** is rigid.

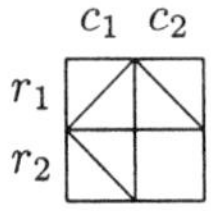

c_1 c_2

r_1

r_2

Figure 8. Grid of **Example 1.**

The system of equations is

$$\begin{aligned} \rho_1 &= 90^\circ \\ \rho_1 &= \kappa_1 + 90^\circ \\ \rho_1 &= \kappa_2 + 90^\circ \\ \rho_2 &= \kappa_1 + 90^\circ, \end{aligned}$$

which is easily solved to yield $\rho_1 = \rho_2 = 90^\circ$ and $\kappa_1 = \kappa_2 = 0^\circ$. Since this is the only solution, the grid is rigid.

1.4 The Degree of Freedom of a Grid

We know from linear algebra that to specify that a system of linear equations has a unique solution, the number of independent equations must equal the number of variables. In terms of rigidity, this says that the number of braces required to rigidify a grid is equal to the number of angles that we are free to choose; this number, in effect, is the degree to which the unbraced grid is free to move.

To draw a deformation of a grid, we need only to choose directions for the posts and beams. We have, leaving the top-left beam fixed, $m + n - 1$ free choices to do so, as indicated in **Figure 9**.

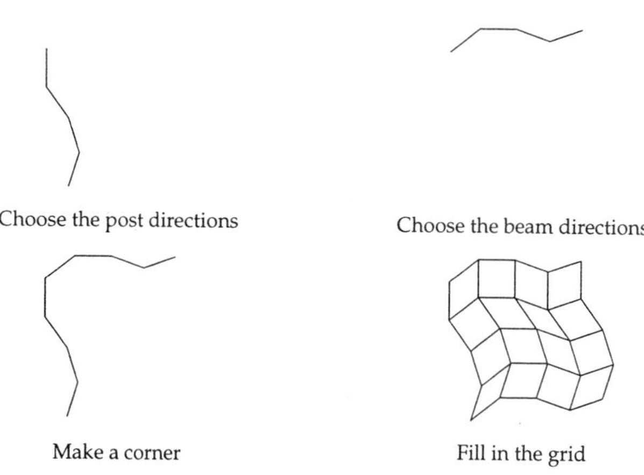

Figure 9. The steps in drawing a deformation of a grid.

For an $n \times m$ grid, there are $(n+m-1)$ variables, $\{\rho_2, \ldots, \rho_n\}$ and $\{\kappa_1, \ldots, \kappa_m\}$, since we are assuming that $\rho_1 = 90°$. So, to specify a unique solution, we require $(n + m - 1)$ independent equations, and hence at least $(n + m - 1)$ braces.

The *degree of freedom* of a grid is the number of braces required to rigidify it, so the degree of freedom of an $n \times m$ grid is $(n + m - 1)$.

Of course, not every choice of $(n + m - 1)$ braces will rigidify a grid, since, as is not surprising, some placements of braces do more work than others. See **Figure 10**; these grids each have four braces, but one is rigid and the other deforms.

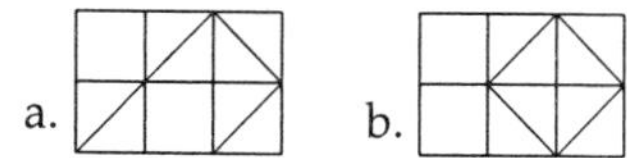

Figure 10. a. An efficient bracing. **b.** A wasteful bracing.

Exercises

1. Write down the system of angle equations for each of the grids in **Figure 10**. Show that the first system has only the trivial solution. Find a nontrivial solution for the second system.

2. Determine which of the following grids are rigid.

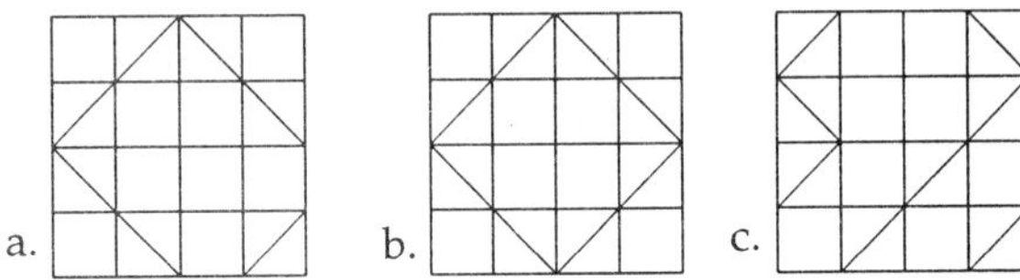

If a grid is nonrigid, then draw a deformation.

3. Find nine braces that make a 5×5 grid rigid.

4. Suppose in a grid that there is some brace that is the only brace in its row, and that same brace is the only brace in its column. Prove that the grid has a deformation.

2. Graphs and Grids

2.1 The Brace Graph

Mathematically, a grid is a collection of rows of parallel posts, together with columns of parallel beams. If there is a brace in square (i, j), then the posts in row i are perpendicular to the beams in column j. So the braces in a grid induce a binary relation on the set of rows and columns, and this relation is naturally represented by a graph.

The *brace graph* of a grid has vertices $\{r_1, r_2, r_3, \ldots\}$, one for each row, as well as vertices $\{c_1, c_2, c_3, \ldots\}$, one for each column, and an edge between r_i and c_j if there is a brace in square (i, j).

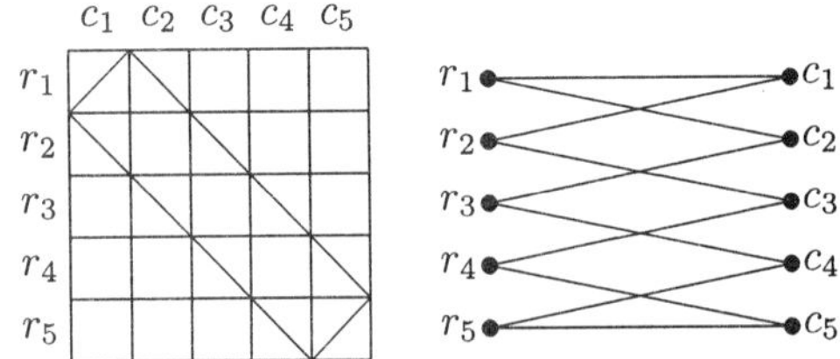

Figure 11. A grid (left) and its brace graph (right).

Figure 11 gives the brace graph for a 5×5 grid. Note that when we draw a graph, each vertex is indicated by a solid dot. Do not confuse any crossing of edges in the picture with vertices. (In fact, this graph can be redrawn so that there are no edge crossings.)

The brace graph shows which squares are braced, so the graph carries all of the information that is contained in the picture of the braced grid. Hence we should be able to study the rigidity of braced grids directly from their brace graphs.

All brace graphs are *bipartite*, that is, the vertex set can be partitioned into two subsets so that all the edges join a vertex in the first set with a vertex in the second. This means that not every graph can be the brace graph of a grid. However, every bipartite graph is the brace graph of a grid.

Exercises

5. Prove that the brace graph in **Figure 11** can be redrawn as a decagon.

6. Draw the brace graph for each of the following grids. See if you can find a "nice" picture for each graph.

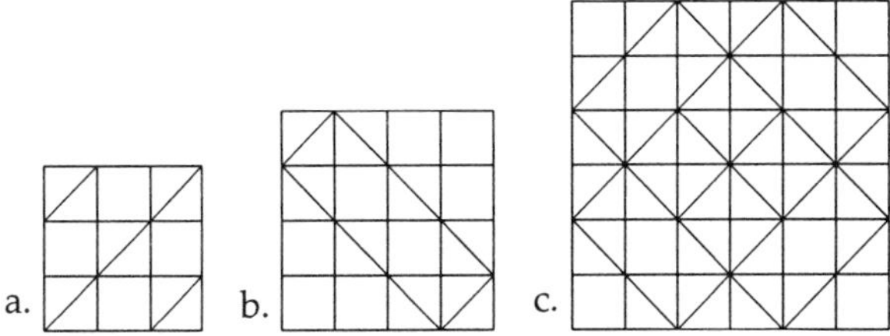

7. Draw the braced grids that go with the following brace graphs.

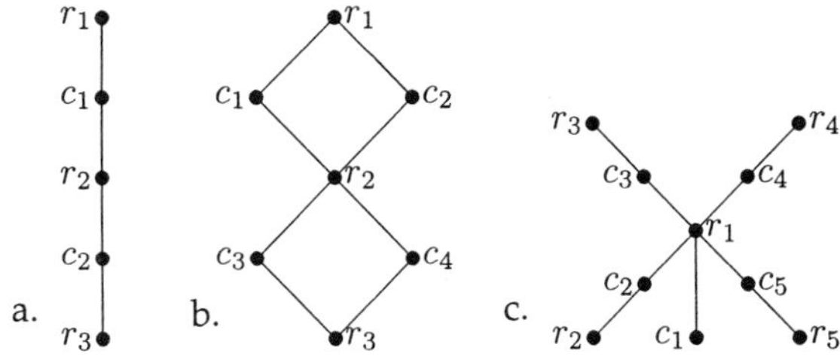

8. Show that the following graphs cannot be the brace graphs of any grid.

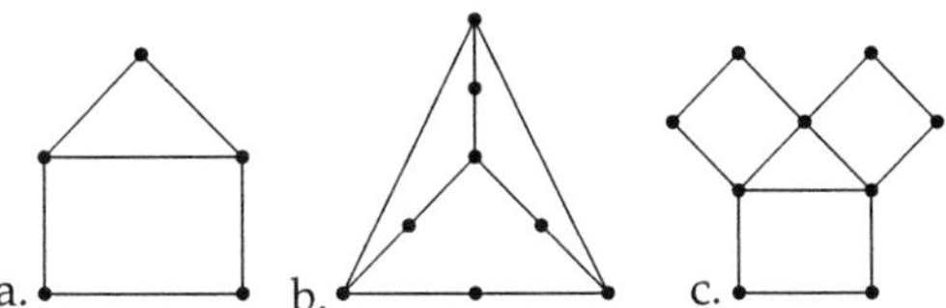

2.2 Connected Brace Graphs

Since each edge in the brace graph corresponds to an angle equation, we can easily detect rigidity via the brace graph.

Theorem. *A braced grid is rigid if and only if its brace graph is connected.*

If a brace graph is *connected*, that is, if there is a path between every pair of vertices, then there is a path from r_1 to every other r_i and c_j. Since each edge in a path stands for the post-beam relation of being perpendicular, a path from r_1 to r_i means that the posts in rows 1 and i are parallel, and a path from r_1 to c_j means that the posts in row 1 are perpendicular to the beams in column j.

We can see this in the case of a 2 × 2 grid with three braces,

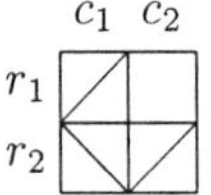

whose brace graph is a path:

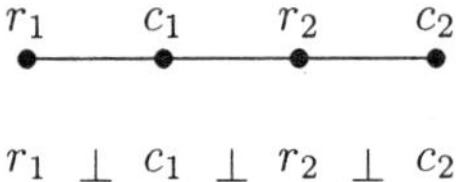

$$r_1 \perp c_1 \perp r_2 \perp c_2$$

Suppose, on the other hand, that the brace graph is *disconnected*, like that of the brace graph in **Figure 12**.

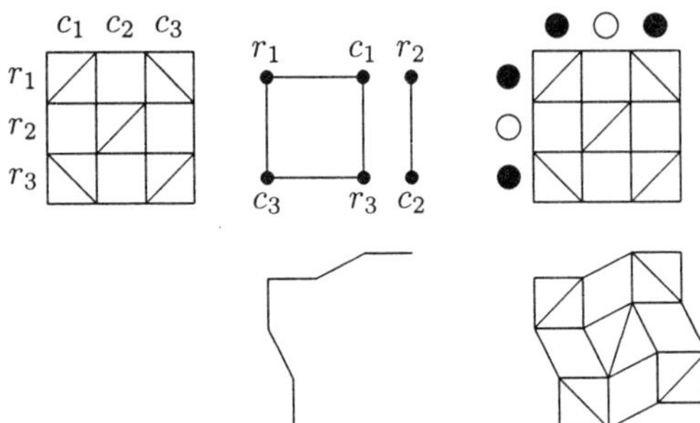

Figure 12. A grid with a disconnected brace graph.

We would like to show that such a braced grid has a deformation. Place a black dot by every row and column that is connected to r_1 in the brace graph. Place a white dot next to all the other rows and columns. Now we have to choose directions for the beams and the posts in the deformation. The directions of the rows of posts and columns of beams with black dots should be perpendicular with each other, since there may be braces between them. The same must be true for the the rows and columns with white dots, since they also may have braces between them. There is no other restriction on the directions, since there is no brace connecting a black-dotted row with a white-dotted column, or connecting a white-dotted row with a black-dotted column. Choose the direction of the posts in the black rows to be vertical (90°) and the posts in the white rows to be at, say, 100°. Then the beams in the black columns must be horizontal (0°), and the beams in the white columns must be at an angle of 10°. This describes our deformation.

2.3 An Algorithm for Detecting Connectivity

If a graph is not connected, we say that it is *disconnected*. It is possible to draw the graph so that it is obvious that it is disconnected simply by drawing the vertices in clumps called *connected components*, as in **Figure 13**.

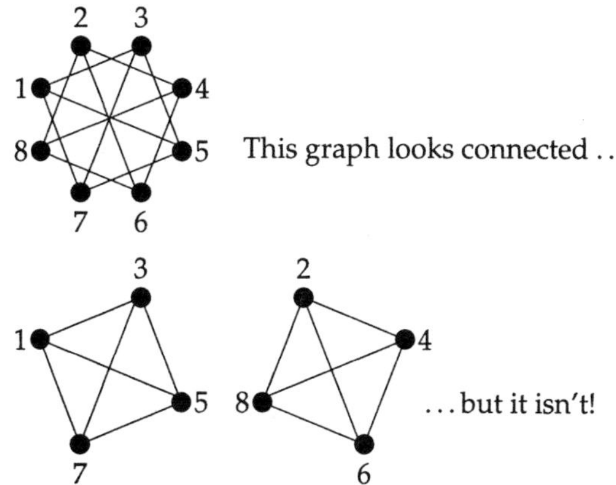

Figure 13. Two views of one graph.

So, while deciding if a graph is connected or not is often quite easy by looking at the picture, we have to be careful that we do not mistake disconnected graphs for connected graphs.

The following algorithm could have been inspired by hauling in a fishing net hand over hand, dropping into the boat what we already have. If we don't get the whole net into the boat this way, then the net is broken.

Algorithm to Determine If a Graph Is Disconnected

Step 0 Pick any vertex, circle it, and call it the active vertex.

Step 1 Circle all the vertices connected by an edge to the active vertex. Put a cross on the active vertex.

Step 2 If all the vertices are circled, then declare the graph connected and STOP.

Step 3 If all the circled vertices have crosses, then declare the graph to be disconnected and STOP.

Step 4 Declare one of the uncrossed circled vertices to be the active vertex and GOTO Step 1.

Exercises

9. Determine which of the following braced grids are rigid.

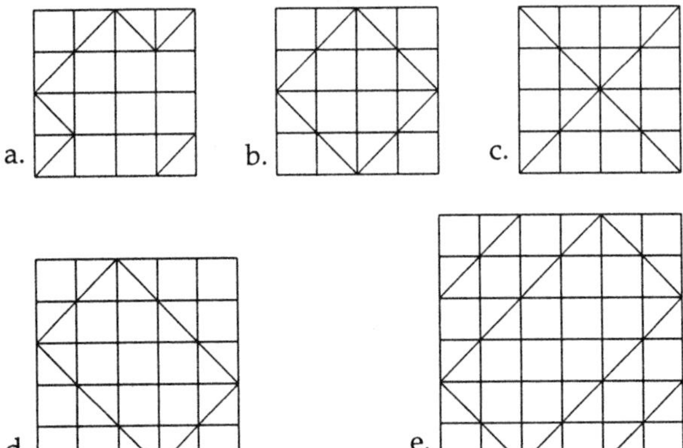

Try to find a deformation for the nonrigid ones. For the rigid ones try to find any unnecessary braces.

10. Explain why none of the following "checkerboard" braced grids is rigid.

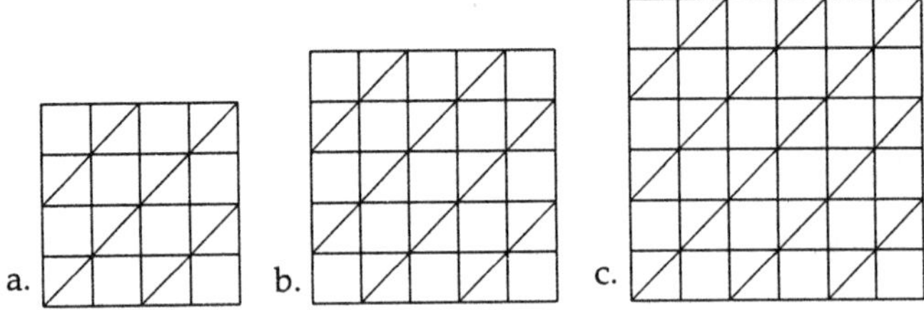

11. The grid below needs just one more brace to make it rigid. Find all the squares in which you can add a brace to make the grid rigid.

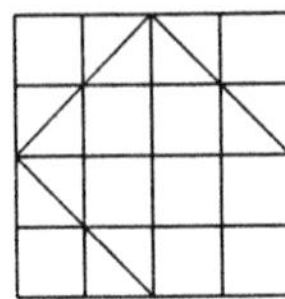

12. Suppose that an $n \times n$ grid is braced so that there are at least two braces in every row and at least two braces in every column. Is it true that the grid is rigid?

2.4 Trees and Efficiency

A *cycle* in a graph is a path that starts and stops on the same vertex. If a graph is connected but has no cycles, it is called a *tree*. In a tree, there is exactly one path between any two vertices. Removing any edge from a tree disconnects it.

If we want to brace a grid efficiently, so that there are no unnecessary braces, we want the brace graph to be connected but have no unnecessary edges. Thus, a grid is efficiently braced when the brace graph is a tree.

Every connected graph has a spanning tree, that is, a subgraph that is a tree and contains all the vertices of the original graph. This means that every rigid braced graph can be made efficient simply by deleting some of the braces.

2.5 Fault-Tolerant Bracings

One problem with efficient bracings is that there is no margin for error. If any brace breaks in an efficiently braced grid, then there is a deformation. In the rigid braced grid of **Figure 14**, there is one brace that is *critical*, that is, if it breaks, then the grid deforms.

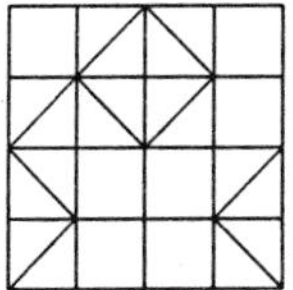

Figure 14. Which is the critical brace?

Such a brace corresponds to a *bridge* in the brace graph, i.e., an edge whose deletion disconnects the graph. The brace graph of **Figure 14** is drawn in **Figure 15**, in which it is clear that the only bridge is the edge between c_1 and r_2.

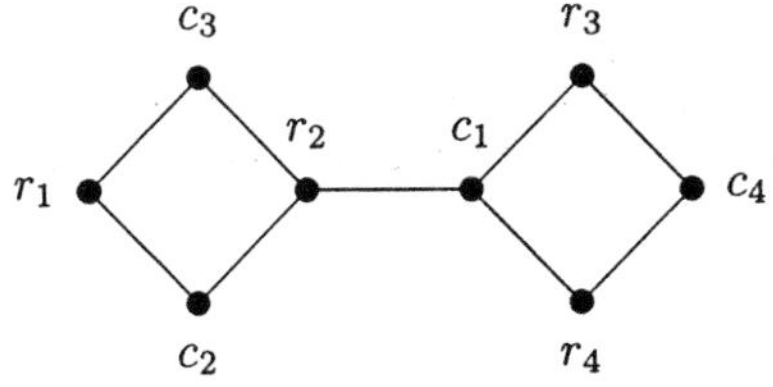

Figure 15. The brace graph for the grid of **Figure 14**.

If we want the grid to be *fault-tolerant*, that is, to remain rigid after the failure of any brace, then the brace graph must be edge 2-connected, that is, the graph must be connected and have no bridges. One way to check 2-connectivity is to make sure that every edge is contained in a cycle. In general, the edge connectivity of the brace graph is related to the *fault tolerance* of the the grid.

2.6 Amazing Transformations

In the sequence of grids of **Figure 16**, each grid is obtained from the previous one by tearing off a column from the right-hand side and gluing it on the left.

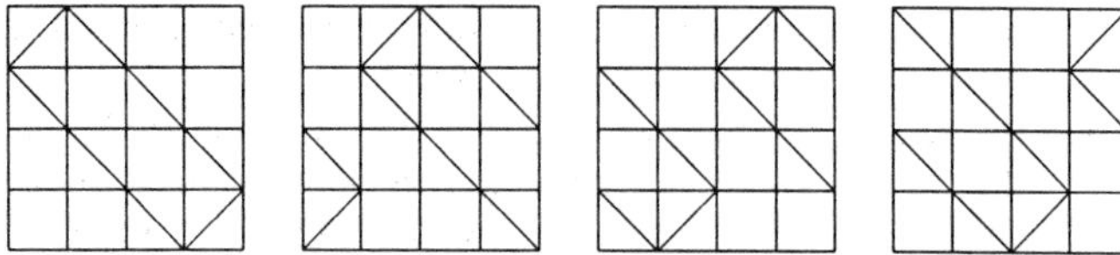

Figure 16. Forming a new grid by tearing a column off the right and gluing it on the left.

Rearranging the rows or columns of a grid simply relabels the vertices in the brace graph, and so can have no effect on the rigidity of the grid.

On the other hand, any automorphism of a connected brace graph corresponds to a rearrangement of the rows and columns of a grid, together perhaps with turning the grid 90°, i.e., interchanging rows and columns.

Exercises

13. Show that there is no fault-tolerant bracing of a $1 \times n$ grid.

14. Find the critical brace in the grid below.

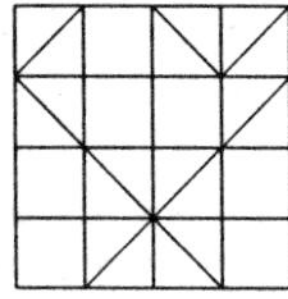

Show how to move exactly one brace so that the resulting grid has no critical braces.

15. Show that the $n \times n$ grid needs only $2n$ braces in a rigid fault-tolerant bracing. What is the form of the brace graph in this case? What about a 3×4 grid? What about 3×5?

16. What is the smallest number of braces that must be added to the grid below so that no brace failure gives a nonrigid grid?

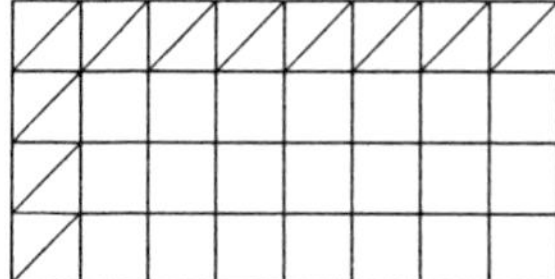

17. Is every tree the brace graph of some grid? Can you find a relation between the shape of the tree and the dimension of the grid? For instance, what grids correspond to stars? To paths?

3. Directed Graphs and Tensegrity

3.1 Bracing Grids with Cables

So far, we have been bracing our grids with rigid rods. A cheaper and lighter solution is to brace the grid with *cables*. This is a different mathematical problem. If we brace a single square with one cable, the square is *not* rigid, since the cable may buckle (see **Figure 17**).

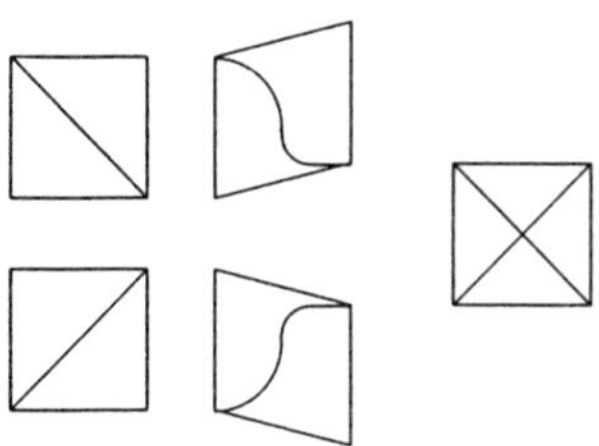

Figure 17. Bracing a single square with one and two cables.

The cable does not stretch, however, so that the corners of the square at which the cable is attached cannot be acute. This means that, unlike rod braces, *it matters which diagonal of the square is braced with a cable*; so let us distinguish these braces as N-cable braces and Z-cable braces.

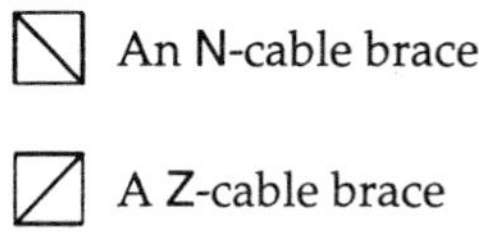

A square with both a Z-cable and an N-cable brace is rigid. This means that we can turn any rigid rod-braced grid into a rigid cable-braced grid by replacing every rod brace with two cables in the same square (see **Figure 18**).

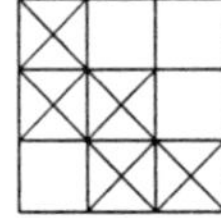 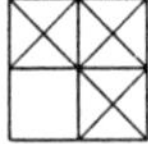 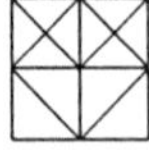

Figure 18. Some rigid cable-braced grids.

In fact, if there were no unnecessary rods, it is not too difficult to verify that every cable is necessary as well—if we remove any one of them, the grid deforms. This is not the only configuration, however. In fact, as we will see shortly, the cable-braced grid on the right of **Figure 18** is also rigid.

Since the degree of freedom of a grid is $(n+m-1)$, we might expect that it will take at least $2(n+m-1)$ cable braces to make an $n \times m$ grid rigid. What is quite startling is that this is often false. There are rigid $n \times n$ cable-braced grids with only $2n$ cables, one more than the number of rods required, even though each cable does only half the work of a rod!

The 2×2 grid can be rigidified with only 4 cables (see **Figure 19**).

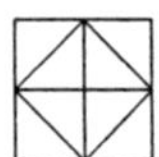 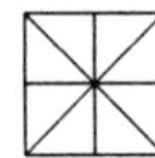 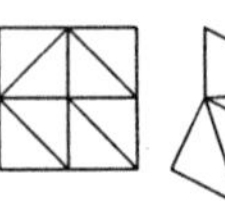

Figure 19. Rigidifying a 2×2 grid.

The cable-braced grid at the left of **Figure 19** is rigid because the cables prevent each of the four angles that meet in the middle from being obtuse. Since they add to 360°, they must all be right angles. The same reasoning applies to the second cable-braced grid, where none of the middle angles can be acute. Notice that the cabling is working together. If one of the cables is attached the other way, as in the third grid, the effect is lost and there can be a severe deformation.

3.2 Cables and Linear Inequalities

Recall that ρ_i is the angle made with the horizontal by the posts in row i, and that κ_j is the angle made with the horizontal by the beams in column j (see **Figure 20**), with $70° \leq \rho_i \leq 110°$ and $-20° \leq \kappa_j \leq 20°$. As before, we assume that $\rho_1 = 90°$.

If square (i, j) has an N-cable brace, then $\rho_i - \kappa_j \leq 90°$; that is, $\rho_i \leq \kappa_j + 90°$. If square (i, j) has a Z-cable brace, then $\rho_i - \kappa_j \geq 90°$; that is, $\rho_i \geq \kappa_j + 90°$. We illustrate these in **Figure 21**.

For rod braces, we obtained a system of linear equations, so the rigidity could be determined by linear algebra. The use of cable braces gives us instead a system of linear inequalities, so the rigidity of cable-braced grids is a linear

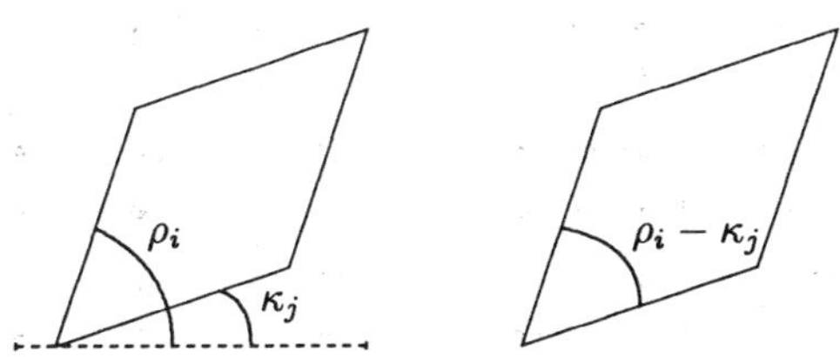

Figure 20. Row and column angles.

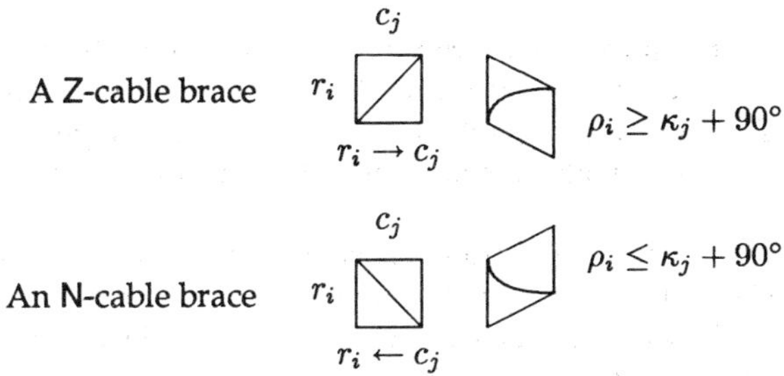

Figure 21. Two kinds of cable braces.

programming problem. As before, however, there is much greater insight to be gained from expressing the problem in terms of graph theory.

3.3 Buckminster Fuller

The word *tensegrity* was coined by R. Buckminster Fuller, a professor at MIT. He also invented the geodesic dome, which may be interpreted as a three-dimensional grid and is considered by some the most significant structural innovation of the twentieth century. The 76-meter-diameter geodesic dome that housed the U.S. pavilion at the World's Fair Montreal in 1976 weighs only about one one-thousandth as much as the 50-meter-diameter dome of St. Peter's Basilica in Rome.

3.4 The Cable-Brace Graph

In the brace graph, an edge between r_i and c_j indicates a brace in square (i, j). For cable braces, we have to distinguish between the N-cable braces and the Z-cable braces. We can indicate this information in the cable-brace graph by directing the edges. We think of directed edges as arrows instead of lines. So if square (i, j) has a Z-cable brace, then we draw an arrow from r_i to c_j. If square (i, j) has an N-cable brace, we draw an arrow from c_j to r_i.

Some cable-brace graphs are drawn in **Figure 22.**

Remember that an ordinary graph is *connected* if you can move from any vertex to any other vertex by moving along the edges. We say that a directed

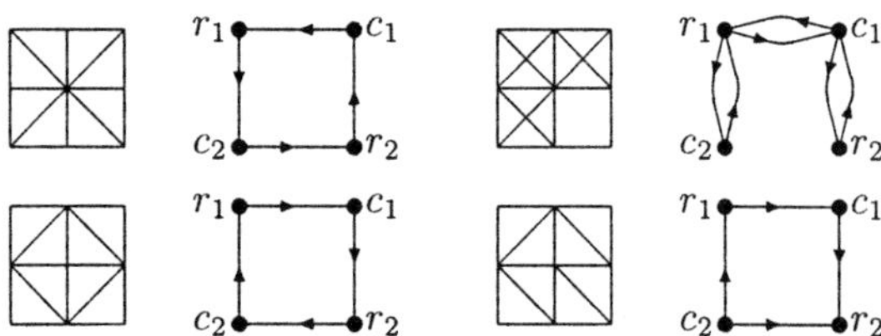

Figure 22. Some cable-brace graphs.

graph is connected if you can move from any vertex to any other vertex by moving along the arrows *in the directions of the arrows*—as if the arrows were one-way streets. The cable-brace graph of the last example in **Figure 22** is not connected, since there is no path from any vertex to c_2, as well as no path from r_2 to any vertex.

The paths in the cable-brace graph are related to the row and column angles. If there is a directed path from r_i to c_j in the cable-brace graph, then $\rho_i \geq \kappa_j + 90°$ in any deformation. To see why, let us look closely at one of the examples from **Figure 22**, which we analyze in **Figure 23**.

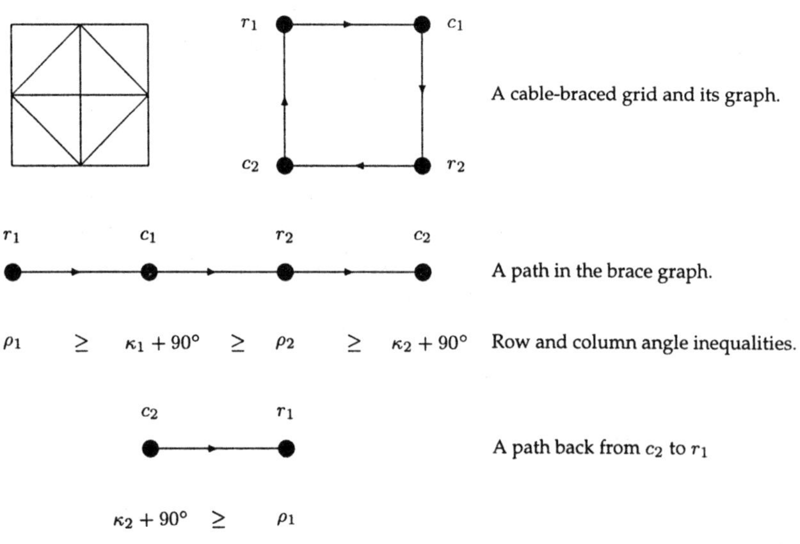

Conclusion: $\rho_1 = \kappa_2 + 90°$ and $r_1 \perp c_2$.

Figure 23. Analysis of the lower-left cable-brace graph of **Figure 21**.

Theorem. *A cable-braced grid is rigid if and only if its cable-brace graph is connected.*

Figure 24 shows an example of a cable-braced grid and its cable-brace graph.

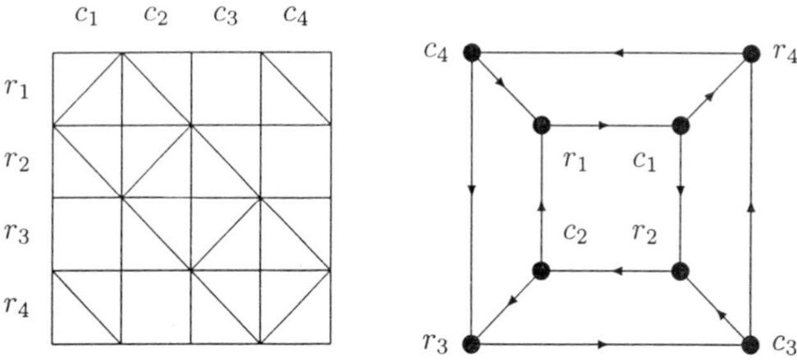

Figure 24. An example of a cable-braced grid and its cable-brace graph.

Let's check that the graph is connected. There is a path from r_1 to any vertex in the inner square, by going around clockwise. Following the arrow from c_1 to the outer square, we can get to every vertex of the outer square. We can get back to r_1 from any vertex on the outer square via the arrow from c_4 to r_1, and we can get to r_1 from any vertex on the inner square by going around clockwise.

3.5 An Algorithm for Directed Graphs

Determining whether or not a directed graph is connected is more difficult than the analogous problem for undirected graphs. One way to do it is to check whether there is a directed path between each pair of vertices. This can be rather tedious for large directed graphs. Here is a somewhat more efficient algorithm:

Algorithm to Determine If a Directed Graph Is Disconnected

Step 0 Pick a vertex v and write $f(v) = 0$.

Step 1 If $f(x) = 0$ for every vertex x, then the directed graph is connected. STOP.

Step 2 If there is no unpicked edge with origin p such that $f(p)$ is maximal, then the directed graph is disconnected. STOP.

Step 3 Choose an unpicked edge e with origin p such that $f(p)$ is maximal. Let q be the terminus of e. If $f(q)$ is undefined, set $f(q) = f(p) + 1$. If $f(q)$ is defined, then redefine every $f(x)$ for which $f(x) > f(q)$ to be $f(q)$. Mark e picked.

Step 4 GOTO Step 1.

Exercises

18. Draw the cable-brace graph for each of the following grids. Check the grids for rigidity.

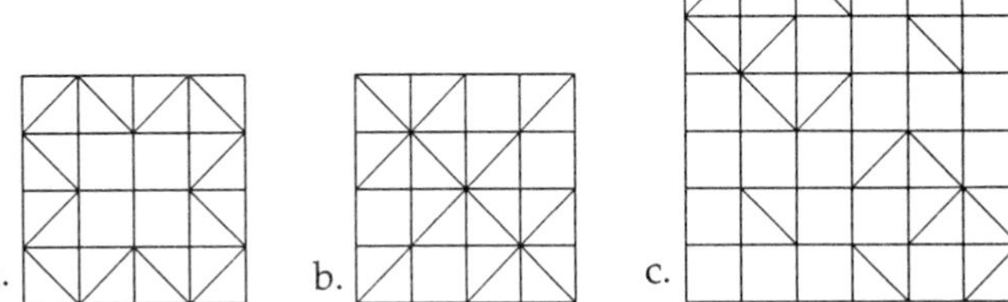

19. The following cable-braced grid is not rigid. Draw the cable-brace graph and see if you can make it rigid by switching some cables from N-cables to Z-cables .

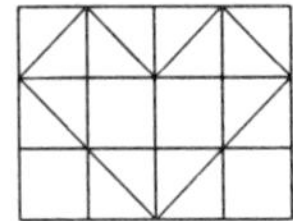

20. Show the following: If a cable-braced grid is rigid, then replacing all the N-cables with Z-cables and vice versa also gives a rigid cable bracing.

21. A 4×4 grid with 8 cables is rigid if and only if the corresponding cable-brace graph is a directed cycle. Why? Below are two examples. How many others can you find? What can you say about them? Can you find all of them?

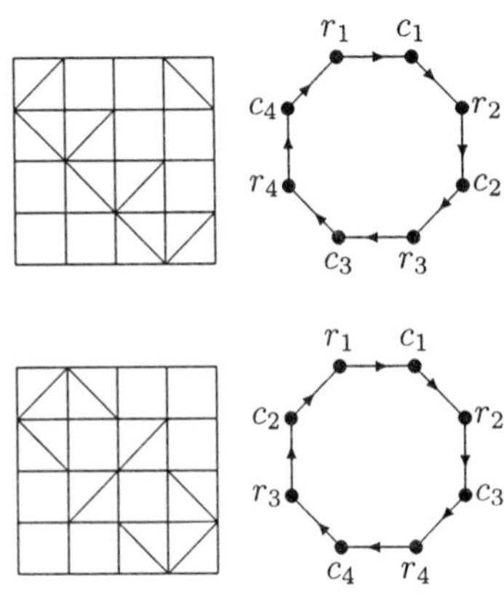

4. Solutions to the Exercises

1. Grid **a** has system $\rho_1 = 90°$, $\rho_1 = \kappa_2 + 90°$, $\rho_1 = \kappa_3 + 90°$, $\rho_2 = \kappa_3 + 90°$, and $\rho_2 = \kappa_1 + 90°$. This system has only the trivial solution, so the grid is rigid.

   The system for the grid **b** is $\rho_1 = \kappa_2 + 90°$, $\rho_1 = \kappa_3 + 90°$, $\rho_2 = \kappa_2 + 90°$, and $\rho_2 = \kappa_3 + 90°$. Since there is no equation for κ_1, we can choose the solution $\rho_1 = \rho_2 = 90°$, $\kappa_1 = 10°$, and $\kappa_2 = \kappa_3 = 0°$, which corresponds to the deformation

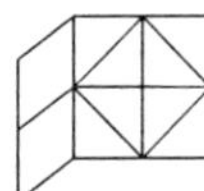

2. Both grids **a** and **c** are rigid. Grid **b** allows the nontrivial solution $\rho_1 = \rho_4 = 90°$, $\rho_2 = \rho_3 = 70°$, $\kappa_2 = \kappa_3 = 0°$, and $\kappa_2 = \kappa_3 = -20°$:

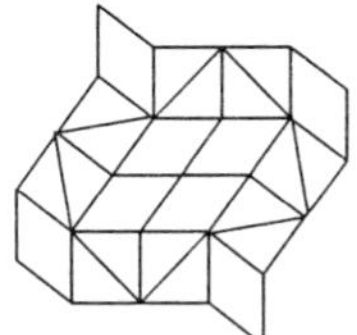

3. Brace every square in the first row and in the first column.

4. Suppose that there is a brace in square (p, q) and no other braces in row p or column q. Then a nontrivial solution to the angle equations is $\rho_i = 90°$, for $i \neq p$; $\rho_p = 100$; $\kappa_j = 0°$, for $j \neq q$; and $\kappa_q = 10°$.

5. The vertices occur around the decagon in the order

$$\{r_1, c_2, r_3, c_4, r_5, c_5, r_4, c_3, r_2, c_1, r_1\}.$$

6.

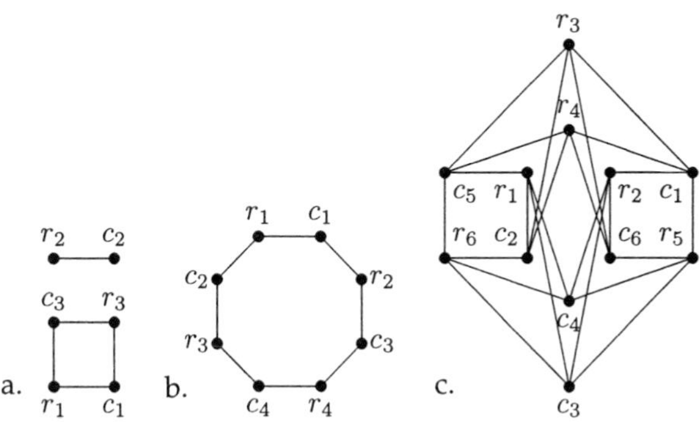

7.

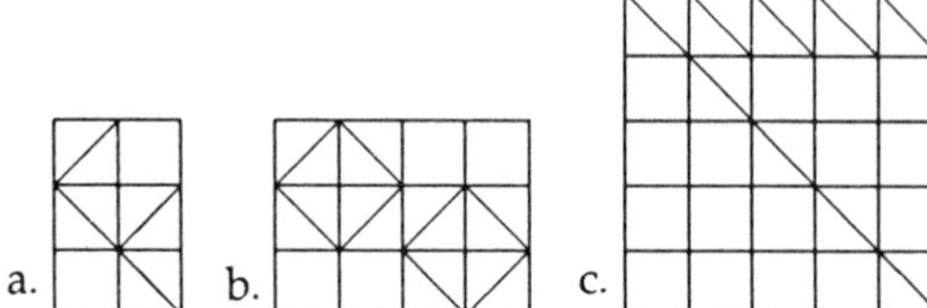

8. A bipartite graph can only have even cycles. Graphs **a** and **c** have cycles of length 3, while graph **b** has a cycle of length 5.

9. Grid **a** is rigid with no unnecessary braces. Grids **b** and **c** deform. The deformation of **b** is shown in the solution to **Exercise 2**. The deformation for **c** is drawn below.

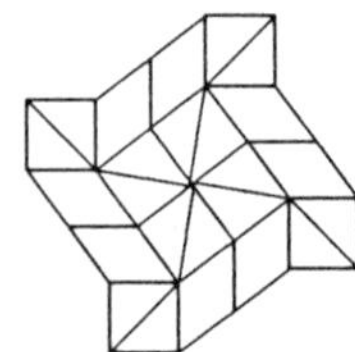

Both grids **d** and **e** are rigid and have unnecessary braces. The graph of **d** is a cycle, and the graph of **e** is given below.

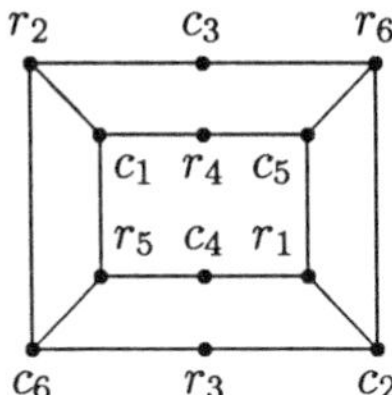

In both cases, deleting any brace leaves a rigid grid.

10. The braces in the checkerboard grids connect only the even rows with the odd columns, and the odd rows with the even columns, so the brace graph is always disconnected even though half the squares are braced.

11. The only squares that *don't* work are $(4, 3)$ and $(3, 4)$.

12. It is not true. This condition means that every vertex in the brace graph must touch at least 2 edges. To find a grid that deforms, we have to find a disconnected graph such that each vertex touches two edges, for instance two disjoint cycles.

13. A $1 \times n$ grid has only one row, so in the brace graph there can only be one edge to each column vertex; thus, a rigid bracing of a $1 \times n$ grid requires every square be braced.

14. The brace graph has one bridge, the edge between r_1 and c_3.

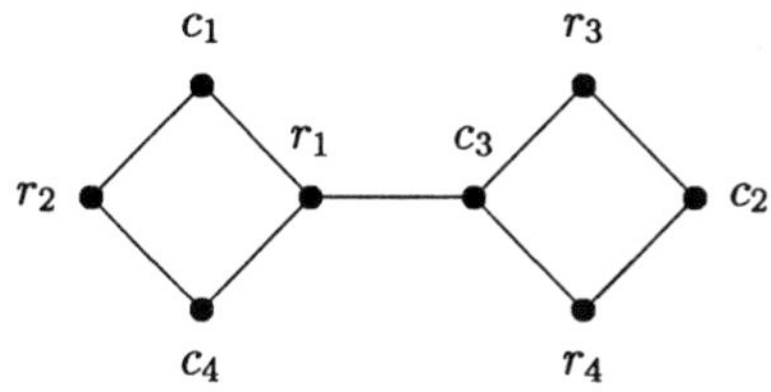

So the brace in square $(1, 3)$ is critical. Moving the brace from $(4, 3)$ to $(4, 4)$, for example, gives a fault-tolerant bracing.

15. This amounts to showing that the only edge 2-connected graph on k vertices and k edges is a cycle.

16. The brace graph is connected with 7 pendant column vertices and 3 pendant row vertices, so a fault-tolerant bracing requires at least 7 new braces. It is easy to find 7 that work.

17. Every tree is the brace graph of a grid. An easy induction shows that every tree is bipartite.

18. All three grids are rigid.

19. If the cables in squares $(3, 2)$ and $(3, 3)$ are switched, then the grid is rigid.

20. Switching the directions of the cables switches the arrows in the brace graph, so the paths from and to r_1 are switched.

21. This amounts to showing that the only connected directed graph on k vertices and k edges is a directed cycle. See **Exercise 15**.

5. Appendix: Notes for the Instructor

The study of grids lends itself very nicely to student investigation using simple materials. It is advisable to start with experiments with models, so that the students gain insight into the complexity of the problem and appreciate the simplicity of the combinatorial solution.

The cheapest and most readily available material for rods is stiff cardboard. A balance must be struck between smaller strips, which are stiffer, and larger strips, which are easier to manipulate. For connectors, flat-headed push pins capped off by refill erasers work quite well. These models work best while laying flat on a table top.

Wooden craft sticks (or tongue depressors) work very well when bolted together. Drilling the holes must be done with care, since the wood splits easily. Use rather long bolts, so that the bolt extends quite a bit beyond the nuts (two thin nuts are best); then the braces can be added over the ends of the bolts without unscrewing the nuts.

Chemistry sticks, or "d-stix," are easy to put together but have the big disadvantage that there is no resistance to crumpling in the third dimension.

A set should contain enough "rods" to make a 5×5 grid, that is, 60 rods, together with 10 to 15 braces.

For cables, string with loops tied at the correct lengths is quite sufficient. (If elastic string is used, then the rigid tensegrity grid will neatly snap back into position when deformed.)

For introducing the students to grids via models, the first question that you want to ask the student is, "How many diagonals are needed?" The next question is, "Where do you place them?" Students usually very quickly find some answer for a particular unbraced grid by experimentation. The first part of this Module lets them check by familiar linear-algebra tools that the answers were correct and also gives them more examples so that they see that there are actually quite a few optimal solutions.

Section 2 of this Module is an elegant translation of the rigidity problem into graph theory. It is a good idea to redo the exercises to **Section 1** here, so that the students realize how much easier the answers are to obtain using the combinatorics. Point out that the exercises to this section would entail great difficulty using only the methods of **Section 1**.

Section 3 illustrates nicely the distinction between directed and undirected graphs. By experimentation again, students should feel the striking difference between the use of rigid versus tension bracings. The placement of every rigid diagonal enlarges a rigid substructure and decreases the degree of freedom. In the case of tension braces, there might not be any rigid substructure at all until the last string miraculously rigidifies the entire grid!

There are always some students who want to go beyond on their own. A quite simple question to ask these students is: If you do not have braces of the length of diagonals of the grid squares, but you do have rods of the same sidelength as the grid squares, how do you achieve rigidity? What shape does your resulting rigid structure have?

A bit more challenging is the following: If some posts or beams of your grid break, how many diagonals do you need now, and where do you place them, to re-rigidify the grid?

A more ambitious project is to examine the case of three-dimensional grids. References to this question are contained in the bibliography.

References

Baglivo, J.A., and J.E. Graver. 1983. *Incidence and Symmetry in Design and Architecture.* New York: Cambridge University Press.

Bolker, E. 1977. Bracing grids of cubes. *Environment and Planning B* 4: 157–172.

________. 1979. Bracing rectangular frameworks II. *SIAM Journal of Applied Mathematics* 36: 491–508.

________, and H. Crapo. 1977. How to brace a one-story building. *Environment and Planning B* 4: 125–152.

________. 1979. Bracing rectangular frameworks. *SIAM Journal of Applied Mathematics* 36: 473–490.

Graver, J., B. Servatius, and H. Servatius. 1993. *Combinatorial Rigidity.* Providence, RI: American Mathematical Society.

Recski, A. 1988/89. Bracing cubic grids—a necessary condition. *Discrete Mathematics* 73: 199–206.

________. 1989. Symmetric bracing of one-story buildings with cables and asymmetric bracings of one-story buildings with rods. In *Symmetry of Structures,* Budapest, vol. 2, 471–472.

________. 1991. One-story buildings as tensegrity frameworks II. *Structural Topology* 17: 43–52.

About the Author

Brigitte Servatius received M.S. degrees in mathematics and physics from the University of Graz, Austria, and a Ph.D. degree in mathematics from Syracuse University. Her main research interests are matroids, especially rigidity matroids, as well as graph and combinatorial group theory. She enjoys coaching the Putnam team, directing undergraduate and graduate research, and using models, puzzles, and games in the classroom. She is coauthor of *Combinatorial Rigidity* published by the AMS in the Graduate Studies in Mathematics series in 1993.

UMAP

Modules in Undergraduate Mathematics and Its Applications

Published in cooperation with the Society for Industrial and Applied Mathematics, the Mathematical Association of America, the National Council of Teachers of Mathematics, the American Mathematical Association of Two-Year Colleges, The Institute of Management Sciences, and the American Statistical Association.

Module 745

Pinochle Numbers

Richard Iltis
Tom Linton

Applications of Combinatorics, Discrete Mathematics, and Probability to Card Games

COMAP, Inc., Suite 210, 57 Bedford Street, Lexington, MA 02173 (617) 862–7878

INTERMODULAR DESCRIPTION SHEET:	UMAP Unit 745
TITLE:	Pinochle Numbers

AUTHOR:	Richard Iltis
	Dept. of Mathematics
	Willamette University
	Salem, Oregon 97301
	`iltis@willamette.edu`
	
	Tom Linton
	Dept. of Mathematics
	Willamette University
	Salem, OR 97301
	and
	Western Oregon State College
	Monmouth, OR 97361
	`linton@fsa.wosc.osshe.edu`

MATHEMATICAL FIELD:	Combinatorics, discrete mathematics, probability
APPLICATION FIELD:	Card games
TARGET AUDIENCE:	Students in the latter part of an introductory course in combinatorics or discrete mathematics.
ABSTRACT:	We use the setting of the card game of pinochle to develop and practice counting techniques, including combinations, generating functions, Pólya's block walking, inclusion–exclusion, and fixing types.
PREREQUISITES:	Understanding of combinations, permutations, binomial coefficients and the binomial theorem, and the principle of inclusion–exclusion. Familiarity with generating functions and with counting combinations by block walking will enrich what is learned from this Module.

COMAP, Inc., Suite 210, 57 Bedford Street, Lexington, MA 02173
(800) 77–COMAP = (800) 772–6627 (617) 862–7878

Pinochle Numbers

Richard Iltis
Dept. of Mathematics
Willamette University
Salem, Oregon 97301
iltis@willamette.edu

Tom Linton
Dept. of Mathematics
Willamette University
Salem, Oregon 97301
and
Western Oregon State College
Monmouth, OR 97361
linton@fsa.wosc.osshe.edu

Table of Contents

MODULES AND MONOGRAPHS IN UNDERGRADUATE MATHEMATICS AND ITS APPLICATIONS (UMAP) PROJECT

The goal of UMAP is to develop, through a community of users and developers, a system of instructional modules in undergraduate mathematics and its applications, to be used to supplement existing courses and from which complete courses may eventually be built.

The Project was guided by a National Advisory Board of mathematicians, scientists, and educators. UMAP was funded by a grant from the National Science Foundation and is now supported by the Consortium for Mathematics and Its Applications (COMAP), Inc., a nonprofit corporation engaged in research and development in mathematics education.

1. Introduction

On a Discrete Mathematics examination, one of us asked the question:

> A pinochle deck comprises two each of the nine, ten, jack, queen, king, and ace from each suit—clubs, diamonds, hearts, and spades—and thus contains 48 cards. In a *hand* of twelve cards, a *run* is a ten, jack, queen, king, and ace of the same suit; a *pinochle* is the pair of queen of spades and jack of diamonds.
>
> 1. How many different hands are possible?
> 2. How many hands contain a run?
> 3. How many hands contain a pinochle?
> 4. What is the probability that a hand with a run in spades also has a pinochle?

Because a pinochle deck has duplicate cards, these were very difficult exam questions. Their solution can involve combinations, generating functions, Pólya's block walking, and inclusion–exclusion. Although they are too difficult for an exam, they are the basis of a good student project.

Let us consider the first one: How many different hands are possible? The order in which the cards are dealt into a hand does not matter, but there are still two ways that we might count the number of hands. First, we could consider the two cards from any pair of twins, say the two queens of diamonds, as indistinguishable. This is an interesting combinatorial problem, but it is not the count that we need to do probability calculations. Second, we could consider the two cards from any twins as distinguishable, perhaps one has a red back and the other blue. This is the approach that is fruitful for calculating probabilities.

2. Counting Distinct Hands

2.1 Combinations

We will look at both ways of counting, beginning with the assumption that twins are identical. We generalize and introduce some notation. Suppose that the objects are two each of n distinct types, and let $\overline{\left[{n \atop k}\right]}$ denote the number of ways to choose k objects without regard to order from the $2n$ objects. We will also call $\overline{\left[{n \atop k}\right]}$ the number of *hands* of size k that can be chosen from n types. The number of distinct pinochle hands would be denoted by $\overline{\left[{24 \atop 12}\right]}$. We first record some ways to calculate the *pinochle numbers* $\overline{\left[{n \atop k}\right]}$ using ordinary combinations.

Fact 1.
$$\overline{\left[{n \atop k}\right]} = \sum_{j=0}^{\lfloor k/2 \rfloor} \binom{n}{j}\binom{n-j}{k-2j}.$$

Proof: For each j from 0 to $\lfloor k/2 \rfloor$, choose j types that will be included as twins, and from the remaining $n-j$ types choose $k-2j$ of them that will be included singly. □

Fact 2.
$$\overline{\begin{bmatrix} n \\ k \end{bmatrix}} = \sum_{j=0}^{\lfloor k/2 \rfloor} \binom{n}{k-j}\binom{k-j}{j} = \sum_{m=\lceil k/2 \rceil}^{k} \binom{n}{m}\binom{m}{k-m}.$$

Proof: For each j from 0 to $\lfloor k/2 \rfloor$, choose $k-j$ types that will be included; then from these types choose j of them that will be included as twins (the other $k-2j$ are included singly). The second equation follows from the first by the change of variable $m = k - j$. □

2.2 Generating Functions

Generating functions give us another method of computing $\overline{\left[{n \atop k}\right]}$. Because we want to choose each of the n types either zero, one, or two times, the generating function is $g(x) = (1 + x + x^2)^n$, and $\overline{\left[{n \atop k}\right]}$ is the coefficient of x^k. Similar to the binomial theorem, we have:

$$(1 + x + x^2)^n = \sum_{k=0}^{2n} \overline{\begin{bmatrix} n \\ k \end{bmatrix}} x^k.$$

We use our generating function to calculate $\overline{\left[{n \atop k}\right]}$.

Fact 3.
$$\overline{\begin{bmatrix} n \\ k \end{bmatrix}} = \sum_{j=0}^{\lfloor k/3 \rfloor} (-1)^j \binom{n}{j}\binom{n+k-1-3j}{k-3j}.$$

Proof: Expand the generating function:

$$\begin{aligned}(1 + x + x^2)^n &= \left(\frac{1-x^3}{1-x}\right)^n = (1-x^3)^n(1-x)^{-n} \\ &= \sum_{s=0}^{n} \binom{n}{s}(-1)^s x^{3s} \cdot \sum_{r=0}^{\infty} \binom{r+n-1}{r} x^r.\end{aligned}$$

Fact 3 then follows from the expansion of the generating function, because the coefficient of x^k in the product

$$\sum_{s=0}^{n} a_s x^{3s} \cdot \sum_{r=0}^{\infty} b_r x^r$$

is

$$\sum_{j=0}^{\lfloor k/3 \rfloor} a_j b_{k-3j}.$$
□

2.3 Inclusion–Exclusion

The form of the sum on the right-hand side of **Fact 3** is suggestive of an inclusion–exclusion argument. If we let $\{1, 2, \ldots, n\}$ denote a set of n types of objects (cards), then $\overline{\left[{n \atop k}\right]}$ is the number of different hands of size k chosen from the deck $\{1, 1, 2, 2, \ldots, n, n\}$. Let x_i denote the number of cards in a hand that are of type i, for $i = 1$ to n, and consider the equation

$$\sum_{i=1}^{n} x_i = k. \tag{1}$$

Then $\overline{\left[{n \atop k}\right]}$ is the number of integer solutions to **(1)** subject to the constraints that $0 \le x_i \le 2$, for $i = 1$ to n.

For inclusion–exclusion arguments, we use notation as in [Grimaldi 1994]. Let c_i denote the condition that $x_i > 2$ (equivalently, $x_i \ge 3$), and let

- N denote the number of solutions to **(1)**;
- $N(c_i)$ denote the number of solutions to **(1)** that satisfy condition c_i (and perhaps more conditions);
- $N(c_i\, c_j)$ the number of solutions to equation **(1)** that satisfy conditions c_i and c_j (and perhaps others); etc.

Then inclusion–exclusion shows that the number of distinct hands is

$$\overline{\left[{n \atop k}\right]} = N - \sum_{i=1}^{n} N(c_i) + \sum_{1 \le i < j \le n} N(c_i\, c_j) - \sum_{1 \le i < j < l \le n} N(c_i\, c_j\, c_l) + \cdots .$$

Typically, $\sum N(c_i)$ is called S_1, $\sum N(c_i\, c_j)$ is S_2, etc. In this case, the sums corresponding to more than $\lfloor \frac{k}{3} \rfloor$ conditions are all zero. As well, we have

- $N = \binom{n+k-1}{k}$;
- each $N(c_i) = \dbinom{n+k-4}{k-3}$;
- each $N(c_i\, c_j) = \dbinom{n+k-7}{k-6}$; and
- as long as $j \le \lfloor k/3 \rfloor$, each $N(c_{i_1}\, c_{i_2}\, \ldots\, c_{i_j}) = \dbinom{n+k-1-3j}{k-3j}$.

Because there are $\binom{n}{j}$ many ways to pick $1 \le c_{i_1} < c_{i_2} < \cdots < c_{i_j} \le n$, we see that

$$\overline{\left[{n \atop k}\right]} = \sum_{j=0}^{\lfloor k/3 \rfloor} (-1)^j \binom{n}{j} \binom{n+k-1-3j}{k-3j}.$$

2.4 How Many Hands?

We can use any of the formulas for $\overline{\left[{n \atop k}\right]}$ to compute the number of distinct pinochle hands as:

$$\begin{aligned}
\overline{\left[{24 \atop 12}\right]} &= \sum_{j=0}^{6} \binom{24}{j}\binom{24-j}{12-2j} && \text{(by \textbf{Fact 1})}\\
&= \sum_{j=0}^{6} \binom{24}{12-j}\binom{12-j}{j} = \sum_{m=6}^{12} \binom{24}{m}\binom{m}{12-m} && \text{(by \textbf{Fact 2})}\\
&= \sum_{j=0}^{4} (-1)^j \binom{24}{j}\binom{24+12-1-3j}{12-3j} && \text{(by \textbf{Fact 3})}\\
&= 287,134,346.
\end{aligned}$$

2.5 Special Hands

Example. One of the combinations that earns meld points in a pinochle hand is an *around*, or four of a kind—cards of the same denomination (ace, king, queen, or jack) in all four suits. For instance, the number of distinct pinochle hands with an around in aces (at least one ace from each suit) is given by:

$$\sum_{j=0}^{4} \binom{4}{j} \overline{\left[{20 \atop 8-j}\right]} = \sum_{i=0}^{4} \binom{20}{i}\binom{24-i}{8-2i} = 4{,}388{,}506.$$

In the first sum, we choose the suits in which the aces will appear as twins, and then we choose the remaining cards from the non-aces. Verification of the second sum is an exercise.

This result generalizes:

Fact 4. *Suppose there are n types of cards. The number of hands of size k that contain at least one each of j distinct types is given by:*

$$\sum_{i=0}^{\min\{j,k-j\}} \binom{j}{i} \overline{\left[{n-j \atop k-j-i}\right]} = \sum_{i=0}^{\min\{j,k-j\}} \binom{n-j}{i}\binom{n-i}{k-j-2i}.$$

2.6 The Pinochle Triangle

Next we record some elementary facts about the pinochle numbers $\overline{\left[n \atop k \right]}$ and write out a triangle similar to Pascal's triangle.

$$\overline{\begin{bmatrix} n \\ k \end{bmatrix}} = \overline{\begin{bmatrix} n \\ 2n-k \end{bmatrix}} \qquad \text{(symmetry)} \tag{2}$$

$$\overline{\begin{bmatrix} n \\ 1 \end{bmatrix}} = \binom{n}{1}, \qquad \overline{\begin{bmatrix} n \\ 2 \end{bmatrix}} = \binom{n}{1} + \binom{n}{2} = \binom{n+1}{2} \tag{3}$$

$$\overline{\begin{bmatrix} n \\ k \end{bmatrix}} = \overline{\begin{bmatrix} n-1 \\ k \end{bmatrix}} + \overline{\begin{bmatrix} n-1 \\ k-1 \end{bmatrix}} + \overline{\begin{bmatrix} n-1 \\ k-2 \end{bmatrix}} \qquad \text{(recursion)} \tag{4}$$

Statements **(2)** and **(4)** are analogous to the symmetry relation $\binom{n}{k} = \binom{n}{n-k}$ and the recursion relation $\binom{n}{k} = \binom{n-1}{k} + \binom{n-1}{k-1}$ for binomial coefficients. The right-hand side of **(4)** can be thought of combinatorially as: Identify one of the types as special; choose zero, one, or two of the special type; then choose the rest of the hand from the remaining $n-1$ types. The first few lines of a triangle of pinochle numbers $\overline{\left[n \atop k \right]}$ are shown in **Table 1**.

Table 1.
The pinochle triangle.

$n\backslash k$	0	1	2	3	4	5	6	7	8	9	10
0	1										
1	1	1	1								
2	1	2	3	2	1						
3	1	3	6	7	6	3	1				
4	1	4	10	16	19	16	10	4	1		
5	1	5	15	30	45	51	45	30	15	5	1
6	1	6	21	50	90	126	141	126	90	50	...
7	1	7	28	77	161	266	357	393	357	266	...

Note how **(2)**, **(3)**, and **(4)** are illustrated in the triangle. Using

$$(1+x+x^2)^n = \sum_{k=0}^{2n} \overline{\begin{bmatrix} n \\ k \end{bmatrix}} x^k$$

with $x = 1$ and $x = -1$, we obtain

$$3^n = \sum_{k=0}^{2n} \overline{\begin{bmatrix} n \\ k \end{bmatrix}}, \tag{5}$$

$$1 = \sum_{k=0}^{2n} (-1)^k \overline{\begin{bmatrix} n \\ k \end{bmatrix}}. \tag{6}$$

These identities too are illustrated in the triangle; look at sums of the rows and alternating sums of the rows. Equation **(5)** can also be verified by counting the number of hands of sizes from $k = 0$ to $k = 2n$. We have three choices for each type—choose zero, one, or two of that type. Adding **(5)** and **(6)** gives

$$3^n + 1 = 2 \cdot \overline{\sum_{k=0}^{n} \begin{bmatrix} n \\ 2k \end{bmatrix}},$$

which says that doubling the sum of the even-numbered entries in a row gives one more than a power of three.

2.7 Block Walking

Besides algebraic and combinatorial methods of proof, Pólya's idea of block walking (see Tucker [1989]) often can be used to discover and verify identities involving binomial coefficients.

A walker begins at corner $(0,0)$ and proceeds downward, turning left or right (as viewed by us) at each corner. A corner is labeled (n,k) if it is n blocks from $(0,0)$ and is reached with k right-hand turns. The number of routes from corner $(0,0)$ to corner (n,k) is $\binom{n}{k}$ (see **Figure 1**).

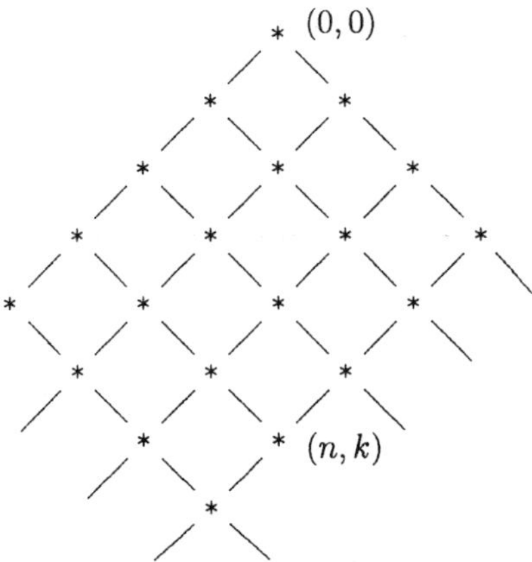

Figure 1. Block walking, with labeling by binomial coefficients.

To extend block walking to pinochle numbers, we also allow the walker an additional option at each corner of proceeding directly down. A corner is labeled (n,k) if it is n blocks from $(0,0)$ and is reached with a turn count of k, where a left turn counts 0, straight down counts 1, and a right turn counts 2. The number of routes from corner $(0,0)$ to corner (n,k) is $\overline{\left[{n \atop k}\right]}$ (see **Figure 2**).

2.8 Three Summation Identities

We now offer three summation identities. We begin with the basic recursion $\overline{\left[{n \atop k}\right]} = \overline{\left[{n-1 \atop k}\right]} + \overline{\left[{n-1 \atop k-1}\right]} + \overline{\left[{n-1 \atop k-2}\right]}$, then repeatedly replace the leftmost, middle, or rightmost term by again using the basic recursion.

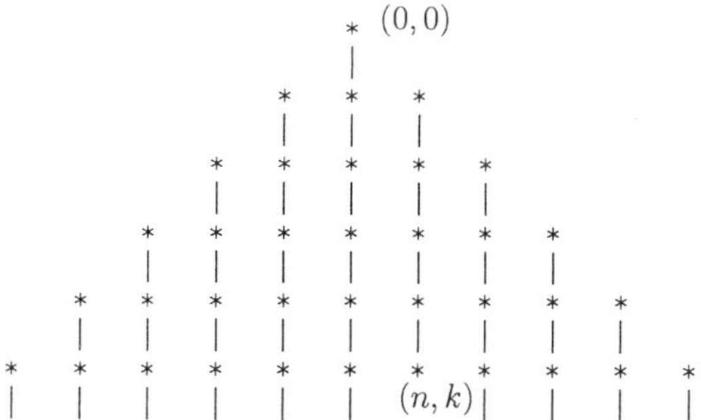

Figure 2. Block walking, with labeling by pinochle numbers.

Fact 5. *Each identity can be viewed as following a walker back to the boundary of the triangle of blocks from a given corner.*

The first identity follows the walker back up and to the right. There are two cases, corresponding to the hand size being odd or even, which can be combined as one identity:

$$\overline{\begin{bmatrix} m+k+1 \\ 2m+1 \end{bmatrix}} = \sum_{j=0}^{k} \left\{ \overline{\begin{bmatrix} m+j \\ 2m \end{bmatrix}} + \overline{\begin{bmatrix} m+j \\ 2m-1 \end{bmatrix}} \right\}$$

$$\overline{\begin{bmatrix} m+k+1 \\ 2m+2 \end{bmatrix}} = \sum_{j=0}^{k} \left\{ \overline{\begin{bmatrix} m+j \\ 2m+1 \end{bmatrix}} + \overline{\begin{bmatrix} m+j \\ 2m \end{bmatrix}} \right\}$$

$$\overline{\begin{bmatrix} \lfloor p/2 \rfloor + k + 1 \\ p+1 \end{bmatrix}} = \sum_{j=0}^{k} \left\{ \overline{\begin{bmatrix} \lfloor p/2 \rfloor + j \\ p \end{bmatrix}} + \overline{\begin{bmatrix} \lfloor p/2 \rfloor + j \\ p-1 \end{bmatrix}} \right\}. \quad \textbf{(7)}$$

The second identity follows the walker back up and to the left. There are two cases, corresponding to the hand size being odd or even:

$$\overline{\begin{bmatrix} m+k+1 \\ 2k+1 \end{bmatrix}} = \sum_{j=0}^{k} \left\{ \overline{\begin{bmatrix} m+j \\ 2j \end{bmatrix}} + \overline{\begin{bmatrix} m+j \\ 2j+1 \end{bmatrix}} \right\}$$

$$\overline{\begin{bmatrix} m+k+1 \\ 2k \end{bmatrix}} = \sum_{j=0}^{k} \left\{ \overline{\begin{bmatrix} m+j \\ 2j-1 \end{bmatrix}} + \overline{\begin{bmatrix} m+j \\ 2j \end{bmatrix}} \right\}. \quad \textbf{(8)}$$

The third identity follows the walker back up vertically:

$$\overline{\begin{bmatrix} m+1 \\ k \end{bmatrix}} = \sum_{j=0}^{k} \left\{ \overline{\begin{bmatrix} m-k+j \\ j-2 \end{bmatrix}} + \overline{\begin{bmatrix} m-k+j \\ j \end{bmatrix}} \right\}. \quad \textbf{(9)}$$

Proof: Identity **(7)** follows by repeatedly replacing the rightmost term in the basic recursion relation. It is analogous to the binomial-coefficient identity

Identity **(8)** follows by repeatedly replacing the leftmost term in the basic recursion relation. It is analogous to the binomial-coefficient identity $\sum_{k=0}^{n} \binom{r+k}{k} = \binom{r+n+1}{n}$, which Graham et al. [1994] call *parallel summation*. A view of this identity as block walking is shown in **Figure 4**.

For pinochle numbers, there is a new summation; identity **(9)** follows by repeatedly replacing the middle term in the basic recursion relation. A view of this identity as block walking is shown in **Figure 5**. □

2.9 Pinochle-Number Identities

Completely analogous to Vandermonde's convolution for binomial coefficients [Graham et al. 1994, 169–170],

$$\sum_{k=0}^{r} \binom{m}{k}\binom{n}{r-k} = \binom{m+n}{r},$$

we have:

Fact 6.
$$\sum_{k=0}^{r} \overline{\begin{bmatrix} m \\ k \end{bmatrix}} \cdot \overline{\begin{bmatrix} n \\ r-k \end{bmatrix}} = \overline{\begin{bmatrix} m+n \\ r \end{bmatrix}}$$

Proof: We could use block walking, but instead we argue combinatorially. Suppose that $m+n$ types are divided into two groups of m and n each. For each k from 0 to r, choose k objects from the first group and $r-k$ objects from the second group. □

An identity similar to one listed by Tucker [1989, 207] for binomial coefficients is a corollary to **Fact 6**:

$$\sum_{k=0}^{2n} \overline{\begin{bmatrix} n \\ k \end{bmatrix}}^{2} = \overline{\begin{bmatrix} 2n \\ 2n \end{bmatrix}}.$$

So, for example, the sum of the squares of the elements of the third row in the pinochle number triangle—1, 3, 6, 7, 6, 3, 1—equals $\overline{\left[\begin{smallmatrix} 6 \\ 6 \end{smallmatrix}\right]} = 141$.

Fact 7.
$$\sum_{k=0}^{2n} k \overline{\begin{bmatrix} n \\ k \end{bmatrix}} = n \cdot 3^n$$

Proof: Consider the deck $D = \{1, 1, 2, 2, \ldots, n, n\}$. For a hand $h \subseteq D$, let $|h|$ denote the number of cards in h (for example, if $h = \{0, 0, 1\}$, then $|h| = 3$). The left-hand side of this identity equals

$$\sum_{h \subseteq D} |h|,$$

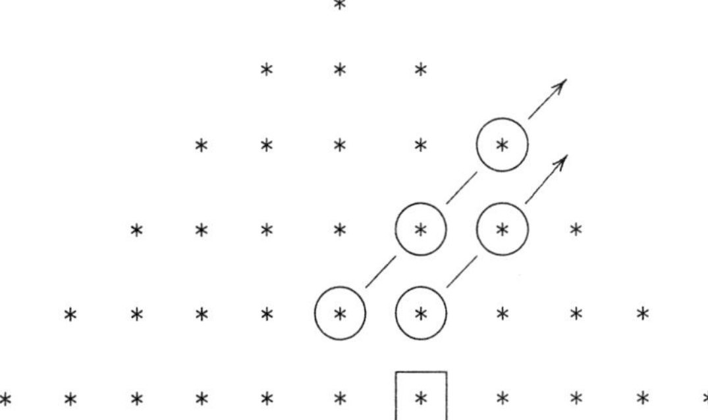

Figure 3. Block-walking view of upper summation, **(7)**.

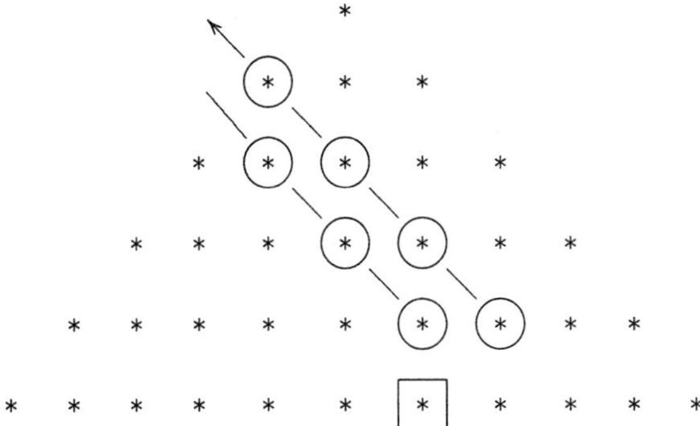

Figure 4. Block-walking view of parallel summation, **(8)**.

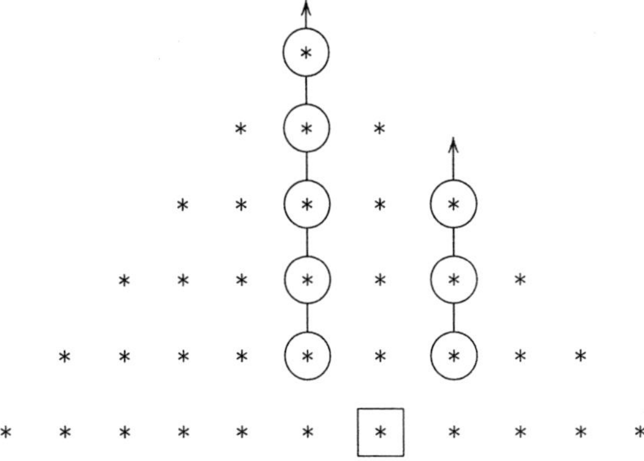

Figure 5. Block-walking view of **(9)**.

because there are $\overline{\left[{n \atop k}\right]}$ distinct hands of size k. As well, by **(5)** on p. 5, $\sum_{k=0}^{2n} \overline{\left[{n \atop k}\right]} = 3^n$, so there are 3^n different hands (of varying sizes) in the deck. The hand $h' = \{1, 2, 3, \ldots, n\}$, consisting of one copy of each of the n distinct cards, is its own complement (the rest of the deck is identical to this hand). All other hands appear in pairs, h and $D \setminus h$. Each of these pairs has a total of $2n$ cards, while h' has n cards. Thus,

$$\sum_{h \subseteq D} |\, h \,| = \frac{3^n - 1}{2} \cdot 2n + n = n \cdot 3^n. \qquad \square$$

Fact 8.

$$\overline{\begin{bmatrix} n+1 \\ a+b+1 \end{bmatrix}} = \sum_{k=0}^{n} \left\{ \overline{\begin{bmatrix} k \\ a \end{bmatrix}} \cdot \overline{\begin{bmatrix} n-k \\ b \end{bmatrix}} + \overline{\begin{bmatrix} k \\ a-1 \end{bmatrix}} \cdot \overline{\begin{bmatrix} n-k \\ b \end{bmatrix}} + \overline{\begin{bmatrix} k \\ a \end{bmatrix}} \cdot \overline{\begin{bmatrix} n-k \\ b-1 \end{bmatrix}} \right\}$$

Proof: This identity is similar to the binomial-coefficient identity

$$\binom{n+1}{a+b+1} = \sum_{k=0}^{n} \binom{k}{a} \binom{n-k}{b}.$$

A combinatorial proof for the binomial-coefficient identity goes as follows. Select any $(a + b + 1)$-element subset of $\{0, 1, 2, 3, \ldots, n\}$. Arrange this subset in increasing order and consider the value of the $(a + 1)^{\text{st}}$ element in this ordered subset. If the $(a + 1)^{\text{st}}$ element is k, then a elements were selected from $\{0, 1, 2, \ldots, k - 1\}$ and b elements from $\{k + 1, k + 2, \ldots, n\}$. There are $\binom{k}{a}$ ways to make the first selection and $\binom{n-k}{b}$ ways to make the second, yielding $\binom{k}{a}\binom{n-k}{b}$ ways to have the $(a + 1)^{\text{st}}$ card equal to k.

Things are a bit more complicated with the pinochle numbers. Because $\overline{\left[{n+1 \atop a+b+1}\right]}$ is the number of $(a + b + 1)$-element hands from $\{0, 0, 1, 1, \ldots, n, n\}$, we may arrange any such hand in increasing order. Then consider the value $k \in [0, n]$ of the $(a + 1)^{\text{st}}$ card. There are three cases to consider.

- *The selected hand contains only one k.*

  We have selected a elements from $\{0, 0, 1, 1, \ldots, k - 1, k - 1\}$ (there are $\overline{\left[{k \atop a}\right]}$ many ways to do this) and b elements from $\{k+1, k+1, k+2, k+2, \ldots, n, n\}$ (there are $\overline{\left[{n-k \atop b}\right]}$ many ways to do this).

- *The ath element in the selected hand is also a k.*

  In this case we selected $a - 1$ elements from $\{0, 0, 1, 1, \ldots, k - 1, k - 1\}$ and b elements from $\{k + 1, k + 1, \ldots, n, n\}$, yielding $\overline{\left[{k \atop a-1}\right]} \cdot \overline{\left[{n-k \atop b}\right]}$ many possibilities.

- *The $(a + 2)^{nd}$ element in the selected hand is also a k.*

  Here we must have chosen a elements from the first k pairs and $b-1$ elements from the last $n - k$ types, yielding $\overline{\left[{k \atop a}\right]} \cdot \overline{\left[{n-k \atop b-1}\right]}$ possibilities for this case. $\square$

Exercises

1. Show that $\overline{\left[{n \atop k}\right]}$ is a kth degree polynomial in n. Note that, for example,

$$\binom{7}{3} = \frac{7(7-1)(7-3)}{3!} \quad \text{can be written as} \quad \left(\frac{1}{6}\right)7^3 - \left(\frac{1}{2}\right)7^2 + \left(\frac{1}{3}\right)7.$$

2. Show that **Fact 2** follows from **Fact 1** by use of properties of binomial coefficients.

3. Verify that $\overline{\left[{24 \atop 12}\right]}$ as given by

$$\sum_{j=0}^{4}(-1)^j\binom{24}{j}\binom{24+12-1-3j}{12-3j}$$

is 287,134,346.

4. Show that

$$\sum_{j=0}^{4}\binom{4}{j}\overline{\left[{20 \atop 8-j}\right]} = \sum_{i=0}^{4}\binom{20}{i}\binom{24-i}{8-2i}.$$

5. Find the number of pinochle hands that contain both an around of kings and a pinochle.

6. Use the triangle of pinochle numbers to find $\overline{\left[{8 \atop 6}\right]}$.

7. Use block walking to prove **Fact 6**.

8. Use generating functions to prove **Fact 7**.

3. Counting for Probabilities

3.1 Probability of a Pinochle

Henceforth, unless otherwise specified, all probabilities and cardinalities refer to a standard pinochle deck and twelve-card hands. To compute the probabilities of certain hands, simply dividing the number (as counted above) of distinct hands with the desired characteristics by the total number of distinct hands is not correct. Some hands occur more often than others. Consider, for example, the hand containing all twelve of the clubs. If we ignore the order in which cards are placed in the hand, there is only one way to obtain this hand—each and every club must be included. Modify this hand by switching one of the nine of clubs with a nine of diamonds, leaving the rest of the hand intact. The second hand will occur more frequently (four times as often) than the first, because the second hand can contain either of the two nines of clubs and either of the two nines of diamonds. We begin by deriving the probability that a hand contains a pinochle (a jack of diamonds and a queen of spades).

3.1.1 Via combinations

We assume that the two copies of any card are distinct and that order does not matter, so there are $\binom{48}{12}$ different hands. We split the deck into two parts: the first part is the two jacks of diamonds and the two queens of spades, and the second is the remaining 44 cards. To have a pinochle, a hand must contain one jack of diamonds and one queen of spades. The remaining ten cards can be any combination from the remaining two cards in the first piece of the deck and the 44 cards in the second piece of the deck.

Fact 9. *The probability that a hand contains a pinochle is given by*

$$Pr(pinochle) = \frac{\sum_{j=0}^{2} \binom{2}{j}\binom{2}{1}^{2-j}\binom{44}{10-j}}{\binom{48}{12}} = \frac{12{,}937{,}981{,}771}{69{,}668{,}534{,}468} = \frac{803}{4324} \approx .185708.$$

Proof: For each j from 0 to 2, choose the number of twins to be included from the first part of the deck. This corresponds to the $\binom{2}{j}$ in the second expression. Then, select one of the two copies of each non–twinned card from the first part of the deck; there are $\binom{2}{1}^{2-j}$ ways to do this. Finally, select the remaining cards from the second part of the deck in any of $\binom{44}{10-j}$ ways. □

Using Mathematica to simulate a pinochle deal, we found that out of 30,000 hands, only 5,494 (about 18.31%) contained a pinochle.

3.1.2 Via permutations

With a more complex computation, we can also derive this probability by assuming that the order in which the cards are received matters, so that the two hands $\{ J\diamondsuit, Q\spadesuit, \dots \}$ and $\{ Q\spadesuit, J\diamondsuit, \dots \}$ are regarded as different. We must select in order twelve cards to complete a hand. Let s_1 denote the index (from 1 to 11) of the first selected card that is a jack of diamonds or a queen of spades. Let s_2 denote the index (from $(s_1 + 1)$ to 12) of the first selection that completes the pinochle. Thus, if the index s_1 corresponds to choosing a queen of spades, then s_2 corresponds to the index of the first jack of diamonds that is selected. Because cards 1 through $(s_1 - 1)$ are not jacks of diamonds or queens of spades, their probabilities of selection are

$$\frac{44}{48}, \frac{43}{47}, \frac{42}{46}, \dots, \frac{44 - s_1 + 2}{48 - s_1 + 2}, \qquad \text{with product} \qquad \frac{P(44, s_1 - 1)}{P(48, s_1 - 1)},$$

where $P(n, k) = \frac{n!}{(n-k)!}$ represents the number of permutations of n distinct objects taken k at a time. The ${s_1}^{\text{st}}$ card can be any of four (two jacks of diamonds and two queens of spades) out of the remaining $(48 - s_1 + 1)$ cards. Once this card is chosen, its twin must be included in the allowed choices for selections s_1+1 to s_2-1. Similar to the previous calculation, these cards will be selected with probability

$$\frac{P(46-s_1, s_2-s_1-1)}{P(48-s_1, s_2-s_1-1)}.$$

The s_2^{nd} card must be one of the two possibilities to complete the pinochle, and the rest of the hand is unrestricted. Thus we obtain the following second derivation for the probability of a pinochle.

Fact 10. *The probability that a hand contains a pinochle is given by*

$$\sum_{s_1=1}^{11} \frac{P(44, s_1-1)}{P(48, s_1-1)} \cdot \frac{4}{49-s_1} \cdot \sum_{s_2=s_1+1}^{12} \frac{P(46-s_1, s_2-s_1-1)}{P(48-s_1, s_2-s_1-1)} \cdot \frac{2}{49-s_2} = \frac{803}{4324}.$$

3.1.3 Via the complement

There are several ways to compute the probability that a hand contains a pinochle by looking at hands that do not contain a pinochle. A hand without a pinochle must be void of jacks of diamonds or void of queens of spades. Using unordered counting and inclusion–exclusion, where

- c_1 is the condition that a hand contains *no* jack of diamonds and
- c_2 is the condition that a hand contains *no* queen of spades,

we can write the probability of a pinochle as

$$\begin{aligned}\Pr(\text{pinochle}) &= \frac{\binom{48}{12} - N(c_1) - N(c_2) + N(c_1\,c_2)}{\binom{48}{12}} \qquad \textbf{(10)}\\ &= \frac{\binom{48}{12} - 2\binom{46}{12} + \binom{44}{12}}{\binom{48}{12}} = \frac{803}{4324}.\end{aligned}$$

3.1.4 Via generating functions

Finally, we employ generating functions. Because each type of card can appear zero, one, or two times (single occurrences can be either of the two copies), the total number of pinochle hands is the coefficient of x^{12} in $(1+2x+x^2)^{24} = (1+x)^{48}$. We demand that at least one jack of diamonds and one queen of spades be in the hand; the number of hands with a pinochle is the coefficient of x^{12} in

$$(1+2x+x^2)^{22}(2x+x^2)^2, \qquad \text{which is} \qquad 4\binom{44}{10} + 4\binom{44}{9} + \binom{44}{8}.$$

This is the same expression as given in **Fact 9**.

3.2 Fixing Types

With the great variety of approaches to calculation, questions about pinochle probabilities are ideally suited for student projects. Our methods generalize to compute numerous probabilities. Above, we fixed two types of cards (jack of diamonds and queen of spades); we found that the number of hands that contain at least one copy of these two types is

$$\sum_{j=0}^{2} \binom{2}{j} \binom{2}{1}^{2-j} \binom{44}{10-j}.$$

We shall refer to this number as $\texttt{fix}\,(2)$, meaning that two types of cards are fixed, or forced to be in the hand. The notion of fixing types arose in the previous section when we established **Fact 4**. The fixed cards must be types; fixing two jacks of diamonds does not correspond to $\texttt{fix}\,(2)$.

Although this notation is not helpful in computing probabilities of some of the common meld values in pinochle (like double pinochles or double marriages), it greatly simplifies the computation of many other probabilities. We let

$\texttt{fix}\,(n)$ be the number of pinochle hands that contain at least one copy of n distinct types of cards,

where n is an integer from 0 to 12. The general formula for $\texttt{fix}\,(n)$ is only slightly more complex than the obvious generalization of the formula for $\texttt{fix}\,(2)$ above. If n is large, one cannot "pair up" all n types in a hand, so the upper limit on the sum must be adjusted. Other than this adjustment, the proof of **Fact 9** goes through word for word here.

Fact 11.
$$\texttt{fix}\,(n) = \sum_{j=0}^{\min\{n,12-n\}} \binom{n}{j} \binom{2}{1}^{n-j} \binom{48-2n}{12-n-j}.$$

One can obtain an alternative formula for $\texttt{fix}\,(n)$ by using inclusion–exclusion and concentrating on the cards not in the hand. Requiring n types of cards to appear at least once in a hand is identical to demanding that both copies of each type *do not* occur in the complement of the hand. A straightforward use of inclusion–exclusion yields:

Fact 12.
$$\texttt{fix}\,(n) = \sum_{j=0}^{n} (-1)^j \binom{n}{j} \binom{48-2j}{36-2j}.$$

Table 2 gives values of $\texttt{fix}\,(n)$ that are relevant for pinochle calculations.

3.3 The Original Exam Questions

The original exam questions can now be answered in terms of these quantities. The first question is ambiguous, but for this part of the paper we are

Table 2.
Values of $\texttt{fix}(n)$ that are relevant for pinochle calculations.

n	$\texttt{fix}(n)$
0	69,668,534,468
1	30,757,916,813
2	12,937,981,771
3	5,150,612,454
4	1,924,545,454
5	667,895,695
6	212,358,985
7	60,738,228
8	15,232,096
9	3,224,576
10	541,952
11	64,512
12	4,096

distinguishing $\binom{48}{12}$ hands. A run fixes five card types and can be in any of four different suits. However, the number of hands with a run is not just $4\,\texttt{fix}(5)$, for this would double-count hands that contain runs in two different suits. A simple use of inclusion–exclusion yields the correct answer of

$$\binom{4}{1}\texttt{fix}(5) - \binom{4}{2}\texttt{fix}(10) = 2{,}668{,}331{,}068. \qquad \textbf{(11)}$$

Thus, the probability of being dealt a run is approximately .0383. See also Gudder [1980].

The answer to the third question, the number of hands that contain a pinochle, is simply $\texttt{fix}(2) = 12{,}937{,}981{,}771$ as illustrated above. The number of hands with a run in spades is $\texttt{fix}(5)$, while the number with a run in spades and a pinochle is $\texttt{fix}(6)$ (the queen of spades can be used both in the run and for the pinochle—the hand is not required to contain both queens of spades). Thus, the probability that a hand with a run in spades also contains a pinochle is given by

$$\frac{\texttt{fix}(6)}{\texttt{fix}(5)} = \frac{42{,}471{,}797}{133{,}579{,}139} \approx .318.$$

3.4 More about the Game of Pinochle

Before calculating some more-challenging probabilities, let us take a moment to describe the game of pinochle in greater detail. Usually the game is played with an even number of players (4, 6, or 8), who are paired as partners. Partners work together to accumulate points via two methods. The first is called *meld points,* wherein certain combinations of cards are assigned numeric values; e.g., a pinochle is worth 40 pts and a run is worth 150 pts. Meld points come in two varieties:

- non-trump melds:
  - **pinochle:** a jack of diamonds and a queen of spades (40 pts);
  - **marriage:** a queen and king of the same suit (20 pts for each marriage);
  - **around:** four of a kind with one of each suit; jacks (40 pts), queens (60 pts), kings (80 pts), or aces (100 pts).
- trump melds:
  - **run:** a jack, queen, king, ten, and ace of trump (150 pts, which includes points for the marriage, so the marriage points are not added to this);
  - **marriage:** a marriage of trump (not in a run) counts 40 pts instead of 20;
  - **nine:** a nine of trump (10 pts each).

Besides these, there are double pinochles (300 pts), double runs (1,500 pts), double marriages (usually worth only 40 pts for non-trump suits and 80 pts in the trump suit), and double arounds (400, 600, 800, and 1,000 pts for jacks, queens, kings, and aces, respectively). Each double meld is simply both copies of each type of card required for the single meld. Trump is declared by the highest bidder in an auction held just after the hand is dealt.

Meld points are counted (and revealed to all players) at the start of the play of the hand. Cards are ranked $9 < J < Q < K < 10 < A$, with trump cards being higher than all non-trump cards. Following rules about when one can play which cards, the players then play out the hand trick by trick. A *trick* consists of each player in turn laying down one card; the highest card takes the trick, and the player winning the trick leads the next trick. Each ten, king, and ace that a player (or partnership) collects in these tricks counts 10 pts. There is also a reward of 10 pts for the player or partnership that takes the final trick. Similar to the *high-card points* in bridge, one can assign a bidding value to a pinochle hand. One rule states that the value of a hand is its meld plus twice the sum of the number of trump cards and aces. A general rule of thumb is to bid up to about the value of your hand. Computing the probability that a hand has a value exceeding, say, 450 is another reasonable question for students to attack on a group basis.

Several versions of pinochle exist, and many have a passing of cards (or collection of extra "kitty" cards) that can greatly improve one's hand. Hence, calculating probabilities from a dealt hand is not completely consistent with the manner in which the game may be played, but nonetheless it is a good source of questions for student projects. Among the group of people whom we play with, there is a special rule about requesting a re-deal. If a player is dealt 5 nines and has no non-trump meld, or 6 (or more) nines (even with non-trump meld), that player may request a re-deal. While the number of hands with 6 or more nines is straightforward to compute,

$$\binom{8}{6}\binom{40}{6} + \binom{8}{7}\binom{40}{5} + \binom{8}{8}\binom{40}{4} = 112{,}830{,}094$$

(so that such a hand occurs with probability about .00162), the number of hands that contain exactly 5 nines and no meld makes an ideal question for a project. The fact that the hand contains 5 nines greatly limits the amount of double counting that one has to deal with. This leaves only 7 cards; so, for example, a hand with a run and an around (plus 5 nines) is impossible. Removing the restriction that the hand has 5 nines makes the question perhaps too ambitious for a project, but the problem could be divided into pieces and solved as a class project.

3.5 More-Challenging Problems

3.5.1 Basic method

It took us several days to calculate the number of hands with no meld; by doing so, one derives many interesting probabilities associated with the game of pinochle. We choose a few illustrative and interesting examples, including one very tedious calculation (so readers can see why we think this is too ambitious for an undivided project), and give tables summarizing our findings and a simulation done on Mathematica.

The basic method of our solution is using the terms `fix` (n) (which implies that probabilities involving double melds will be lost) together with inclusion–exclusion. One can simplify things slightly by realizing that a hand with no meld is equivalent to a hand with

- no pinochle,
- no around, and
- no marriage (a hand cannot have a run if it doesn't already have a marriage). However, probabilities involving runs are then lost, so we add the last case,
- no run.

We ignore nines of trump here, since we do not yet know what the trump suit is.

We let

- c_1 be the condition that a hand contains a pinochle,
- c_2 means that the hand contains an around,
- c_3 implies that the hand has a run, and
- c_4 means that the hand has a marriage.

This saves effort compared to defining a separate condition for a marriage in each suit, a run in each suit, and each type of around, and it also makes the computation a bit more interesting.

Several smaller counting problems arise along the way. If one uses the latter method of defining more conditions, the computations are simpler (but much more numerous). For example, the number of hands with a run in spades, a pinochle, and an around in jacks is simply `fix` (8). It is much more difficult to calculate the number of hands with a run, a pinochle, and an around. Although this typically is done by considering individual cases such as the one above, usually several cases can be lumped together.

The number $\overline{N}$ of hands that satisfy none of our conditions can be computed via inclusion–exclusion as

$$\overline{N} = \binom{48}{12} - S_1 + S_2 - S_3 + S_4,$$

where, with the notation used earlier, we have

$$S_1 = \sum_{i=1}^{4} N(c_i), \qquad S_2 = \sum_{1 \le i < j \le 4} N(c_i\, c_j),$$

$$S_3 = \sum_{1 \le i < j < k \le 4} N(c_i\, c_j\, c_k), \qquad S_4 = N(c_1\, c_2\, c_3\, c_4).$$

We now proceed to calculate several examples by inclusion–exclusion.

3.5.2 A pinochle plus an around

Our first example is a computation of $N(c_1\, c_2)$, the number of hands that contain (at least) a pinochle and an around.

Fact 13. *The number of twelve-card pinochle hands that contain a pinochle and an around is*

$$2\,\texttt{fix}\,(5) + 2\,\texttt{fix}\,(6) - \texttt{fix}\,(8) - 4\,\texttt{fix}\,(9) - \texttt{fix}\,(10) + 2\,\texttt{fix}\,(12) = 1{,}731{,}845{,}200,$$

yielding a probability of about .0249.

Proof: Here, the condition p means that the hand contains a pinochle, a_1 means an around of jacks, a_2 means an around of queens, a_3 means an around of kings, and a_4 means an around of aces. The desired number can be expressed as $S_1 - S_2 + S_3 - S_4$, where S_n denotes the number of hands that contain a pinochle and n arounds, for example,

$$S_1 = \sum_{i=1}^{4} N(p\, a_i) \quad \text{and} \quad S_2 = \sum_{1 \le j < k \le 4} N(p\, c_i\, c_j).$$

Note that $S_4 = 0$, because one cannot have four arounds with only twelve cards. The computation of S_1 breaks into two cases, depending on whether the around is jacks or queens *or* the around is kings or aces. If the around consists of 4 jacks or 4 queens, one of these cards doubles in the pinochle, meaning we are fixing 5 cards. Otherwise, we are fixing 6 cards. Thus,

$$S_1 = 2\,\texttt{fix}\,(5) + 2\,\texttt{fix}\,(6) = 1{,}760{,}509{,}360.$$

The computation of S_2 involves three cases, depending on whether the two arounds consist of two, one, or none of jacks or queens. A hand with a pinochle, an around in jacks, and an around in queens fixes 8 cards. A non-pinochle around (kings or aces) plus a pinochle around (jacks or queens), coupled with a pinochle, fixes 9 cards, while two non-pinochle arounds and a pinochle will fix 10 cards. Hence,

$$S_2 = \texttt{fix}\,(8) + 4\,\texttt{fix}\,(9) + \texttt{fix}\,(10) = 28{,}672{,}352.$$

A hand with three arounds will already contain 12 cards, so jacks and queens must be included in the three types of arounds in order to have also a pinochle. Thus,

$$S_3 = 2\,\texttt{fix}\,(12) = 8{,}192.$$

□

3.5.3 A pinochle plus a marriage

The problem of computing $N(c_1\, c_4)$, the number of hands that contain at least a pinochle and a marriage, does not split into too many subcases; and using our conditions (as opposed to one for each specific meld) saves a fair amount of effort. We carry out this computation in a manner analogous to the last one.

Fact 14. *The number of hands that contain a pinochle and a marriage is*

$$\texttt{fix}\,(3) + 3\,\texttt{fix}\,(4) - 3\,\texttt{fix}\,(5) - 3\,\texttt{fix}\,(6) + 3\,\texttt{fix}\,(7) + \texttt{fix}\,(8) - \texttt{fix}\,(9) = 8{,}480{,}931{,}556,$$

yielding a probability of about .1217.

Proof: If the marriage is in spades, there are 3 fixed types of cards; otherwise, there are 4 cards fixed. Thus $S_1 = \texttt{fix}\,(3) + 3\,\texttt{fix}\,(4)$. We have double-counted all hands with two marriages and thus must subtract them off again. The hands with two marriages and a pinochle fall into two groups: those with a spade marriage, and those with no spade marriage. There are $\binom{3}{1}$ of the first kind, and each of these hands fixes 5 cards, while there are $\binom{3}{2}$ of the second type, each fixing 6 cards. Thus, we should subtract $S_2 = 3\,\texttt{fix}\,(5) + 3\,\texttt{fix}\,(6)$. Now hands with marriages in 3 different suits have been added 3 times in S_1 and subtracted $\binom{3}{2} = 3$ times in S_2. These hands must be added back in. There are $\binom{3}{2}$ ways for a hand with three marriages to contain a marriage in spades (thus fixing 7 cards) and only one way (clubs, diamonds and hearts) to have no spade marriage. This last hand fixes 8 cards, and hence $S_3 = 3\,\texttt{fix}\,(7) + \texttt{fix}\,(8)$. Finally, we should subtract off all hands with a pinochle and a marriage in each suit. Such a hand fixes 9 cards, yielding $\texttt{fix}\,(9)$ as a value for S_4 and completing the computation. □

3.5.4 A pinochle plus an around plus a marriage

So far, our sample calculations have been rather short. As a final example, we sketch a computation that essentially requires considering the conditions for each individual marriage and each individual around. We shall calculate $N(c_1\, c_2\, c_4)$, the number of hands with a pinochle, an around, and a marriage.

Fact 15. *The number of hands that contain a pinochle, an around, and a marriage is*

$$6\,\mathtt{fix}\,(6) - 2\,\mathtt{fix}\,(7) + 4\,\mathtt{fix}\,(8) - 13\,\mathtt{fix}\,(9) + 3\,\mathtt{fix}\,(11) + \mathtt{fix}\,(12) = 1{,}171{,}883{,}982,$$

yielding a probability of about .0168.

Proof: Given that the answer is not so complicated, there is no doubt a better way to compute this quantity. However, the interplay between the cards in each of the meld sequences is high here; and for this reason we shall simply provide tables of the cases with their corresponding numbers. We will compute successively $S_1, S_2, \ldots, S_6$, where

$$N(c_1\, c_2\, c_4) = S_1 - S_2 + S_3 - S_4 + S_5 - S_6.$$

The meaning of each $S_1, S_2, \ldots$ should be clear from the tabulation. Each S_n splits into cases. Each hand being counted must contain a pinochle, one or more arounds, and one or more marriages. Because every hand has a pinochle, we will drop it from our tables. We list the cases in **Tables 3–7** in abbreviated notation, where J, Q, K, and A represent respectively an around in jacks, queens, kings, or aces, and where C, D, H, and S stand for a marriage in clubs, diamonds, hearts, or spades, respectively. Remember that the two pinochle cards must be counted in the last column.

Because three arounds exhausts all 12 cards in a hand, the only nonzero contribution to S_6 is to have J, Q, and K around with all four marriages, making

$$S_6 = \mathtt{fix}\,(12).$$ □

3.5.5 Further computations

From these tables, it is straightforward to calculate such quantities as the number of hands that contain a pinochle, two arounds, and three marriages. The tables actually are ordered so that all hands with a arounds and m marriages appear together (also, each hand covered by S_n will have $a+m = n+1$). Before summarizing the computation of the remaining probabilities, we show how some other sorts of pinochle-hand counting can be carried out by using terms like $\mathtt{fix}\,(n)$.

One common version of the game has two partnerships of two players each. After the highest bidder declares trump, the bidder's partner passes four cards to the bidder. The bidder keeps the best twelve cards from the sixteen and passes four unwanted cards back to the partner. As shown in **(11)** on p. 15, the

Table 3.

$S_1 = 6\,\texttt{fix}\,(6) + 7\,\texttt{fix}\,(7) + 3\,\texttt{fix}\,(8)$.

around(s)	marriage(s)	contribution
A	C, D, or H	$3\,\texttt{fix}\,(8)$
A	S	$1\,\texttt{fix}\,(7)$
K or J	C, D, or H	$6\,\texttt{fix}\,(7)$
K or J	S	$2\,\texttt{fix}\,(6)$
Q	any suit	$4\,\texttt{fix}\,(6)$

Table 4.

$S_2 = 9\,\texttt{fix}\,(7) + 6\,\texttt{fix}\,(8) + 15\,\texttt{fix}\,(9) + 12\,\texttt{fix}\,(10) + 6\,\texttt{fix}\,(11)$

around(s)	marriage(s)	contribution
Q and (J or K)	any one suit	$-8\,\texttt{fix}\,(9)$
Q and A	any one suit	$-4\,\texttt{fix}\,(10)$
J and K	S	$-1\,\texttt{fix}\,(9)$
J and K	C, D, or H	$-3\,\texttt{fix}\,(10)$
A and (J or K)	S	$-2\,\texttt{fix}\,(10)$
A and (J or K)	C, D, or H	$-6\,\texttt{fix}\,(11)$
J	S and (C, D, or H)	$-3\,\texttt{fix}\,(8)$
J	any two from C, D, H	$-3\,\texttt{fix}\,(9)$
Q	any 2 suits	$-6\,\texttt{fix}\,(7)$
K	S and (C, D, or H)	$-3\,\texttt{fix}\,(7)$
K	any two from C, D, H	$-3\,\texttt{fix}\,(8)$
A	S and (C, D, or H)	$-3\,\texttt{fix}\,(9)$
A	any two from C, D, H	$-3\,\texttt{fix}\,(10)$

Table 5.

$S_3 = 7\,\texttt{fix}\,(8) + 7\,\texttt{fix}\,(9) + 12\,\texttt{fix}\,(10) + 16\,\texttt{fix}\,(11) + 11\,\texttt{fix}\,(12)$

around(s)	marriage(s)	contribution
J, Q, and K	any one suit	$4\,\texttt{fix}\,(12)$
any other 3	any one suit	0
A and J	any two from C, D, H	0
A and J	S and (C, D, or H)	$3\,\texttt{fix}\,(12)$
K and Q	any two	$6\,\texttt{fix}\,(9)$
Q and J	any two	$6\,\texttt{fix}\,(10)$
J and K	S and (C, D, or H)	$3\,\texttt{fix}\,(10)$
J and K	any two from C, D, H	$3\,\texttt{fix}\,(11)$
Q and A	any two	$6\,\texttt{fix}\,(11)$
K and A	S and (C, D, or H)	$3\,\texttt{fix}\,(11)$
K and A	any two from C, D, H	$3\,\texttt{fix}\,(12)$
J	C, D, and H	$1\,\texttt{fix}\,(11)$
J	any other 3	$3\,\texttt{fix}\,(10)$
Q	any three	$4\,\texttt{fix}\,(8)$
K	C, D, and H	$1\,\texttt{fix}\,(9)$
K	any other 3	$3\,\texttt{fix}\,(8)$
A	C, D, and H	$1\,\texttt{fix}\,(12)$
A	any other 3	$3\,\texttt{fix}\,(11)$

Table 6.

$S_4 = 6\,\texttt{fix}\,(9) + 7\,\texttt{fix}\,(11) + 15\,\texttt{fix}\,(12)$

around(s)	marriage(s)	contribution
any 4	any one	0
J, Q, and K	any two	$-6\,\texttt{fix}\,(12)$
any other 3	any two	0
Q and K	any three	$-4\,\texttt{fix}\,(9)$
Q and A	any three	$-4\,\texttt{fix}\,(12)$
Q and J	any three	$-4\,\texttt{fix}\,(11)$
J and A	any three	0
J and K	C, D, and H	$-1\,\texttt{fix}\,(12)$
J and K	any other 3	$-3\,\texttt{fix}\,(11)$
K and A	C, D, and H	0
K and A	any other 3	$-3\,\texttt{fix}\,(12)$
J	C, D, H, and S	$-1\,\texttt{fix}\,(12)$
Q or K	C, D, H, and S	$-2\,\texttt{fix}\,(9)$
A	C, D, H, and S	0

Table 7.

$S_5 = \texttt{fix}\,(9) + 6\,\texttt{fix}\,(12)$

around(s)	marriage(s)	contribution
all four	any two	0
J, Q, and K	any three	$3\,\texttt{fix}\,(12)$
any other three	any three	0
J and (Q or K)	all four	$2\,\texttt{fix}\,(12)$
Q and K	all four	$1\,\texttt{fix}\,(9)$
A and (J, Q, or K)	all four	0

probability of being dealt a run is less than .04. Since the bidder is presumed to have a long trump suit and there are only twelve cards in each suit, it is very unlikely that the bidder's partner has more than four trumps to pass. It will almost never be the case that the passer has at least one of each of the five run cards in the trump suit. Hence, if we assume (as most players do) that the passer will give the bidder one of each of the trump run cards that the passer holds, there is almost no loss of generality in assuming that the only way for the bidding team *not* to get a run is for the opposing team to hold both copies of one of the trump run cards. We can then calculate the probability of getting a run after passing.

One method is to use inclusion–exclusion with the conditions c_i (for $i = 1, \ldots, 5$) meaning that the opposition's 24 cards include two jacks (queens, kings, tens, or aces) of trumps. This method gives

$$\overline{N} = \sum_{j=0}^{5} (-1)^j \binom{5}{j} \binom{48-2j}{24-2j} = 7{,}173{,}761{,}357{,}700$$

hands, out of $\binom{48}{24}$, where the opponents do not block the run. Thus, the probability of a run after passing jumps to about .22, and the effects of passing a few cards becomes evident.

A second approach is to group the bidder's and the bidder's partner's hands into one hand (as explained above, this is almost legitimate) and use a version of fix(5) for 24-card hands. Again, assuming the suit of the run is fixed, we are demanding that 5 types of cards appear in a hand of 24 cards. The number of such hands is

$$\texttt{fix}\,(5,24) = \sum_{j=0}^{5} \binom{5}{j} 2^{5-j} \binom{48-2\times 5}{24-5-j} = 7{,}173{,}761{,}357{,}700.$$

The number of hands that contain double melds or double melds with additional meld values can be computed by another modification of $\texttt{fix}\,(n)$. As an illustration, consider the number of hands with a double pinochle (two jacks of diamonds and two queens of spades) and aces around (a very good hand to bid on!). Because we are demanding that the hand has a double pinochle and at least four aces, we are actually fixing 4 types of cards (the aces in each suit) amongst 22 types (the two *pinochle types* have been removed). The number of such hands is then

$$\sum_{j=0}^{4} \binom{4}{j} 2^{4-j} \binom{44-8}{4-j} = 1{,}186{,}369,$$

and the probability of such a hand is about 17 in a million!

Table 8 summarizes our calculations of the number of hands that contain various meld sequences.

Table 8.
The numbers, probabilities, and frequencies in simulation of particular kinds of hands.
P = pinochle, A = around, R = run, and M = marriage.

meld included				Symbolic value	Numerical value	Prob	Freq
P	A	R	M				
1				f_2	12,937,981,771	.186	.183
	1			$4f_4 - 6f_8 + 4f_{12}$	7,606,805,624	.109	.109
		1		$4f_5 - 6f_{10}$	2,668,331,068	.038	.039
			1	$4f_2 - 6f_4 + 4f_6 - f_8$	41,038,858,204	.589	.592
1	1			$2f_5 + 2f_6 - f_8 - 4f_9 - f_{10} + 2f_{12}$	1,731,845,200	.025	.025
1		1		$2f_6 + 2f_7 - f_{10} - 4f_{11} - f_{12}$	545,390,330	.008	.009
1			1	$f_3 + 3f_4 - 3f_5 - 3f_6 + 3f_7 + f_8 - f_9$	8,477,706,980	.122	.121
	1	1		$16f_8 - 24f_{11} - 24f_{12}$	242,066,944	.003	.003
	1		1	$\left\{ \begin{array}{c} 8f_5 - 4f_6 + 8f_7 - 15f_8 \\ -16f_9 + 28f_{10} - 16f_{11} + 10f_{12} \end{array} \right\}$	4,713,744,212	.068	.067
		1	1	$12f_7 - 12f_9 + 4f_{11} - 12f_{12}$	690,372,720	.010	.010
1	1	1		$4f_{10} + 2f_8 + 10f_9 - 8f_{11} - 22f_{12}$	64,271,552	.001	.001
1	1		1	$6f_6 - 2f_7 + 4f_8 - 13f_9 + 3f_{11} + f_{12}$	1,171,883,982	.017	.017
1		1	1	$f_7 + 7f_8 + 2f_9 - 8f_{10} - f_{11} - f_{12}$	169,407,828	.002	.003
	1	1	1	$24f_9 + 4f_{11} - 72f_{12}$	77,352,960	.001	.001
1	1	1	1	$4f_9 + 18f_{10} + f_{11} - 28f_{12}$	22,603,264	.000	.000
no meld				**Exercise 18**	21,836,301,279	.313	.312

Because these computations were carried out using inclusion–exclusion, the listed meld sequence(s) represent a minimal meld for the hands. That is, the row corresponding to "run and pinochle" includes all hands that contain a run and a pinochle, but some of these hands also include other melds. The last column is the frequency with which such hands occurred under a Mathematica simulation. We simulated 30,000 pinochle hands and counted the number n containing each given meld sequence. The frequencies are thus of the form $n/30,000$. Using the normal approximation, we would expect 95% of the Mathematica simulations to give results within 1.96σ of the mean.

For a row with both a run and a marriage, it should be understood that the suit of the run and the suit of the marriage are different. The probabilities and frequencies have been rounded to three decimal places. The letters P, A, R, and M stand for pinochle, around, run, and marriage, respectively. The symbolic value column is expressed in terms of `fix`(n), but `fix`(n) has been abbreviated to f_n.

Table 9 lists probabilities and frequencies of the different meld sequences; it might be referred to as a *macro table*, because each of its rows is just a summary of several *micro* tables like those in **Fact 15**. Each row shows

- the simulated number of successes X out of $n = 30,000$ trials;
- the expected number of successes, np;
- the standard deviation of X, given by $\sigma = \sqrt{np(1-p)}$; and
- the normalized deviation of X from the mean, given by $(X - np)/\sigma$.

To list all possibilities for p pinochles, a arounds, r runs, and m marriages is a bit excessive. We summarize some of these computations in **Table 10**. In contrast to **Table 9**, the symbolic values below have not been adjusted for double counting. Thus, for example, a row corresponding to two arounds and one marriage will simply add the values of `fix`(n) that correspond to each of the six possibilities for two arounds, paired with each of the four possible marriages.

Exercises

9. Find the probability that a pinochle hand contains a pinochle by using the complement and ordered counting.

10. Justify **(10)** on p. 13 by computing $N(c_1), N(c_2)$, and $N(c_1\, c_2)$.

11. Verify that the coefficient of x^{12} in $(1 + 2x + x^2)^{22}(2x + x^2)^2$ is

$$4\binom{44}{10} + 4\binom{44}{9} + \binom{44}{8}.$$

Table 9.
Comparison of simulation with expected results for meld sequences.
P = pinochle, A = around, R = run, and M = marriage.

meld included				Simulated	Expected	Standard	Normalized
P	A	R	M	number	number	deviation	deviation
1				5,494	5,571	67	−1.15
	1			3,256	3,276	54	−0.36
		1		1,167	1,149	33	0.54
			1	17,774	17,672	85	1.20
1	1			755	746	27	0.34
1		1		269	235	15	2.24
1			1	3,625	3,650	57	−0.45
	1	1		94	104	10	−1.00
	1		1	1,995	2,030	44	−0.80
		1	1	312	297	17	0.86
1	1	1		32	28	5	0.82
1	1		1	495	505	22	−0.43
1		1	1	85	73	9	1.41
	1	1	1	27	33	6	−1.09
1	1	1	1	10	10	3	0.09
no meld				9,356	9,403	80	−0.58

Table 10.
Comparison of simulation with expected results for some combinations of meld sequences. P = pinochle, A = around, R = run, and M = marriage.

meld included				Symbolic	Prob	Freq
P	A	R	M	(an overestimate)		
	1		1	$8f_5 + 8f_6$	.101	.0982
	1		2	$12f_6 + 12f_8$	.0392	.0365
	2		1	$4f_8 + 16f_9 + 4f_{10}$	.00165	.00167
	2		2	$6f_8 + 24f_{10} + 6f_{12}$	.00150	.0014
	1		3	$8f_7 + 8f_{10}$	.00704	.00603
	3		m	$2\binom{4}{m} f_{12}$	small	0
	1		4	$2f_8 + 2f_{12}$	.000437	.0004
	2		3	$4f_8 + 16f_{11}$	.000889	.0008
	2		4	$f_8 + 4f_{12}$	.000219	.0002
1	1	1		$2f_8 + 10f_9 + 4f_{10}$	.000931	.00107
1	2	1		$8f_{11} + 14f_{12}$	.00000823	0
1	1	2		$8f_{12}$	.000000470	0
1		1	1	$f_7 + 7f_8 + 4f_9$	.00259	.00293
1		2	1	$4f_{12}$	.000000235	0
1		1	2	$2f_9 + 8f_{10} + 2f_{11}$	.000157	.0001
1		1	3	$f_{11} + 3f_{12}$	.00000110	0
1	1	1	1	$4f_9 + 23f_{10} + 17f_{11} + 4f_{12}$	.000380	.000333
1	2	1	1	$3f_{11} + 31f_{12}$	.0000046	0
1	1	2	1	0	0	0
1	1	1	2	$5f_{10} + 17f_{11} + 13f_{12}$	.0000554	0
1	1	1	3	$2f_{11} + 6f_{12}$	.0000022	0
1	2	1	2	$3f_{11} + 9f_{12}$	.00000331	0
1	2	1	3	$f_{11} + 3f_{12}$	.0000011	0

12. Each of the meld sequences below can be computed directly by using `fix` (n) for some value of n. Find the value of n, and the probability of each sequence, using **Table 2** on p. 15 and ignoring "double counting."
    a) Kings around plus a pinochle.
    b) A run in diamonds, queens around, and a pinochle.
    c) A pinochle plus two runs, with one run in spades or diamonds, and the second run in hearts or clubs.
    d) Jacks or aces around plus one marriage.

13. Verify **Fact 12** using inclusion–exclusion, where the condition c_i (for $1 \leq i \leq n$) states that a 36-card hand contains both copies of cards with type i, and N is the total number of 36-card hands in a 48-card deck.

14. Among the "double melds," double pinochles and double marriages are by far the most common. Each of these has two types of cards, and both copies of each must appear in a hand. To compute probabilities for hands including these double melds, a new version of `fix` (n) is needed. Because it involves calculations with double melds, the name `fixdm` comes to mind.
    a) In terms of generating functions, for $k \leq 8$, `fixdm` (k) will be the coefficient of x^8 in $(1 + 2x + x^2)^{22-k}(2x + x^2)^k$. Explain why this is so.
    b) Using an argument similar to the one in the proof of **Fact 9** (or using part **a** above), show that

$$\texttt{fixdm}\,(k) = \sum_{j=0}^{\min\{k,8-k\}} \binom{k}{j} \binom{2}{1}^{k-j} \binom{44-2k}{8-k-j}.$$

    c) Find the number of hands that contain the following melds, using **Table 11**.
        i. A double pinochle.
        ii. A run in clubs and a double marriage in clubs.
        iii. A double pinochle and an around.

15. Use the tables in the proof of **Fact 15** to find the number of hands with a pinochle, an around, and two marriages.

16. Justify the entry in **Table 8** on p. 23 that states that the number of hands with an around and a run is given by $16f_8 - 24f_{11} - 24f_{12}$.

17. Similar to the last problem, show that the number of hands with a run and a marriage (in different suits) is $12f_7 - 12f_9 + 4f_{11} - 12f_{12}$.

18. A hand with no meld is equivalent to a hand with no pinochle, no around, and no marriage (if it has no marriages, it has no runs). The numbers of hands containing each combination of these three melds are given in **Table 8** on p. 23. Use them to fill in the symbolic column for the *no meld* row of the table.

Table 11.
Values of `fixdm` (k) (see **Exercise 14c**).

k	fixdm (k)
0	177,232,627
1	59,202,442
2	18,076,942
3	4,952,635
4	1,186,369
5	239,128
6	38,128
7	4,288
8	256

19. Compute the probability that a hand contains an around in kings and a pinochle. Compute the ratio of the number of distinct hands (as in Section 2 on counting distinct hands) that contain kings around and a pinochle, and the total number of distinct hands (see **Exercises 5** and **12**). Repeat with hands that contain just a pinochle.

4. Solutions to the Exercises

1. From **Fact 1**, $\overline{\left[{n \atop k}\right]} = \sum_{j=0}^{\lfloor k/2 \rfloor} \binom{n}{j}\binom{n-j}{k-2j}$, we see that $\overline{\left[{n \atop k}\right]}$ is a kth-degree polynomial in n because $\binom{m}{r}$ is a rth-degree polynomial in m.

2. Using the right-hand side of the equation in **Fact 1**, we argue that:

$$\begin{aligned} \binom{n}{j}\binom{n-j}{k-2j} &= \binom{n}{j}\binom{n-j}{n-k+j} \quad \text{(symmetry of binomial coefficients)} \\ &= \binom{n}{k-j}\binom{k-j}{j}. \end{aligned}$$

The last equality follows from the binomial coefficient identity

$$\binom{m}{p}\binom{p}{x} = \binom{m}{x}\binom{m-x}{m-p};$$

from m people choose p people, and from those choose x leaders.

3. We calculate:

$$\begin{aligned}
\overline{\begin{bmatrix}24\\12\end{bmatrix}} &= \sum_{j=0}^{4}(-1)^j\binom{24}{j}\binom{24+12-1-3j}{12-3j}\\
&= \binom{24}{0}\binom{35}{12}-\binom{24}{1}\binom{32}{9}+\binom{24}{2}\binom{29}{6}\\
&\qquad -\binom{24}{3}\binom{26}{3}+\binom{24}{4}\binom{23}{0}\\
&= (1)(834{,}451{,}800)-(24)(2{,}804{,}880)+(276)(475{,}020)\\
&\qquad -(2{,}024)(2{,}600)+(10{,}626)(1)\\
&= 287{,}134{,}346.
\end{aligned}$$

4. The second sum can be obtained from the first by using **Fact 1** and interchanging the order of summation.

$$\begin{aligned}
\sum_{j=0}^{4}\binom{4}{j}\overline{\begin{bmatrix}20\\8-j\end{bmatrix}} &= \sum_{j=0}^{4}\binom{4}{j}\sum_{i=0}^{\lfloor(8-j)/2\rfloor}\binom{20}{i}\binom{20-i}{8-j-2i}\\
&= \sum_{i=0}^{4}\binom{20}{i}\sum_{j=0}^{8-2i}\binom{4}{j}\binom{20-i}{8-j-2i}=\sum_{i=0}^{4}\binom{20}{i}\binom{24-i}{8-2i}.
\end{aligned}$$

The last equality follows from the binomial coefficient identity

$$\sum_{k=0}^{r}\binom{m}{k}\binom{n}{r-k}=\binom{m+n}{r}.$$

5. The number of distinct pinochle hands that contain an around of kings and a pinochle is (see **Fact 4**):

$$\sum_{i=0}^{6}\binom{6}{i}\overline{\begin{bmatrix}18\\6-i\end{bmatrix}}=\sum_{i=0}^{6}\binom{18}{i}\binom{24-i}{6-2i}=330{,}145.$$

6. From the basic recursion relation for pinochle numbers, we have:

$$\overline{\begin{bmatrix}8\\6\end{bmatrix}}=\overline{\begin{bmatrix}7\\6\end{bmatrix}}+\overline{\begin{bmatrix}7\\5\end{bmatrix}}+\overline{\begin{bmatrix}7\\4\end{bmatrix}}=357+266+161=784.$$

7. Consider the corner labeled $(m+n, r)$, which is $m+n$ blocks from $(0,0)$ and is reached with a turn count of r. The number of routes from $(0,0)$ to $(m+n, r)$ is $\overline{\left[{m+n \atop r}\right]}$. We can also count the routes to corner $(m+n, r)$ by stopping at intermediate corners (m, k) that are each m blocks from $(0,0)$ but have differing turn counts of k. It is n blocks, with a turn count of $r-k$, from corner (m, k) to corner $(m+n, r)$, so the number of routes from (m, k) to $(m+n, r)$ is $\overline{\left[{n \atop r-k}\right]}$. Therefore, the number of routes to corner $(m+n, r)$ is the sum of $\overline{\left[{m \atop k}\right]} \cdot \overline{\left[{n \atop r-k}\right]}$ over all intermediate corners that are m blocks from $(0,0)$, which is $\sum_{k=0}^{r} \overline{\left[{m \atop k}\right]} \cdot \overline{\left[{n \atop r-k}\right]} = \overline{\left[{m+n \atop r}\right]}$.

8. Differentiate $(1+x+x^2)^n = \sum_{k=0}^{2n} \overline{\left[{n \atop k}\right]} x^k$ and set $x = 1$.

9.

$$\Pr(\text{pinochle}) = 1 - \frac{P(44,12) + 4P(12,1)P(44,11) + 2P(12,2)P(44,10)}{P(48,12)}$$

$$= 1 - \frac{\binom{44}{12} + 4\binom{44}{11} + 2\binom{44}{10}}{\binom{48}{12}} = \frac{803}{4324}.$$

10. $N(c_1) = N(c_2)$ because both represent the number of hands that contain no copies of one type of card. Removing the two copies of one type leaves 46 cards, and we choose 12 of them. $N(c_1\, c_2)$ represents the number of hands with no jacks of diamonds and no queens of spades; this leaves 44 cards, from which we choose 12. Thus, to count the number of hands with a pinochle, we take all $\binom{48}{12}$ different hands, remove the $\binom{46}{12}$ hands that are void of jacks of diamonds and the $\binom{46}{12}$ hands lacking any queens of spades, and put back in the $\binom{44}{12}$ hands with no jacks of diamonds and no queens of spades (because they were removed twice).

11. $(1+2x+x^2)^{22}(2x+x^2)^2 = (1+x)^{44}(x^2)(4+4x+x^2)$. Thus, the coefficient of x^{12} in $(1+2x+x^2)^{22}(2x+x^2)^2$ is equal to the coefficient of x^{10} in $(1+x)^{44}(4+4x+x^2)$, which is $4\binom{44}{10} + 4\binom{44}{9} + \binom{44}{8}$ (by the binomial theorem).

12. The values for $\texttt{fix}\,(n)$ are read from **Table 2** on p. 15 and these numbers are divided by $\texttt{fix}\,(0) = \binom{48}{12} = 69{,}668{,}534{,}468$ to yield the approximate probabilities.

a) $n = 7, p = \frac{\texttt{fix}\,(7)}{\texttt{fix}\,(0)} = .0008718.$

b) $n = 8, p = \frac{\texttt{fix}\,(8)}{\texttt{fix}\,(0)} = .0002186.$

c) $n = 11, p = 4\frac{\texttt{fix}\,(11)}{\texttt{fix}\,(0)} = .000003704.$

d) $n = 6, p = 8\frac{\texttt{fix}\,(6)}{\texttt{fix}\,(0)} = .02439.$

13. $N = \binom{48}{36} = \binom{48}{12}$. All of the $N(c_i)$'s are equal, as are all of the $N(c_i\, c_j)$'s, etc., and thus inclusion–exclusion yields

$$\texttt{fix}\,(n) = N - \binom{n}{1}N(c_1) + \binom{n}{2}N(c_1\,c_2) - \binom{n}{3}N(c_1\,c_2\,c_3) + \cdots$$
$$\cdots + (-1)^n N(c_1\,c_2\,\ldots\,c_n),$$

because there are $\binom{n}{j}$ ways to select j of the conditions to be satisfied simultaneously. Now, $N(c_1\,c_2\,\cdots\,c_j)$ represents the number of 36-card hands that contain both copies of j different types of cards. This puts $2j$ cards in the hand and leaves $48 - 2j$ cards, from which we must select $36 - 2j$ to finish off the hand.

14. **a)** We fix k types that must be included (along with the 4 cards from the double meld), so their generating function factor should be $2x + x^2$. There are $22 - k$ types left over that may be included zero, one, or two times. These have a generating function factor of $1 + 2x + x^2$ (again because there are 2 ways to pick one copy, we use $2x$). Since 4 cards are already designated as in, we need 8 more, explaining why we look for the coefficient of x^8.

b) We pick from zero to $\min\{k, 8-k\}$ of the fixed types to include as doubles. Say we select j of these. The remaining $(k - j)$ fixed types are included as singles, and hence there are 2 ways to pick each of these. Now we have the 4 original doubled cards, $2j$ doubled cards from the k fixed types, and $(k - j)$ single cards from the k fixed types. This leaves $12 - 4 - 2j - k + j = 8 - k - j$ cards left to select. These are to be selected from the $48 - 4 - 2k = 44 - 2k$ remaining cards in the deck.

c) (i) $\texttt{fixdm}\,(0) = 177{,}232{,}627$. (ii) We need to add the ten, jack, and ace of clubs to the double marriage, and $\texttt{fixdm}\,(3) = 4{,}952{,}635$. (iii) If the around is in jacks or queens, we fix an additional three cards. If the around is kings or aces, we fix four additional cards. Thus we get $2\,\texttt{fixdm}\,(3) + 2\,\texttt{fixdm}\,(4) = 12{,}278{,}008$ hands. This, however, double-counts hands that have two arounds. If the two arounds are jacks and queens, we fix six additional cards; if they are (jacks or queens) and (kings or aces), we fix seven additional cards. If the two arounds are kings and aces, we need eight additional cards. Thus, the corrected answer is

$$12{,}278{,}008 - (\,\texttt{fixdm}\,(6) + 4\,\texttt{fixdm}\,(7) + \texttt{fixdm}\,(8)\,) = 12{,}222{,}472.$$

15. The signs $(+\,-)$ in the last columns need to be switched. Then, add the last 7 rows of S_2; subtract all but the first 2 rows of S_3; add all of the rows from S_4; subtract all of S_5; and add S_6. You should end up with

$$9\,\texttt{fix}\,(7) - \texttt{fix}\,(8) + 4\,\texttt{fix}\,(9) - 9\,\texttt{fix}\,(10) - 7\,\texttt{fix}\,(11) + 3\,\texttt{fix}\,(12) =$$
$$538{,}993{,}396.$$

16. Use inclusion–exclusion. Because any around coupled with any run fixes eight cards, $S_1 = 16\,\texttt{fix}\,(8)$. If a hand has two arounds and a run, it will require 11 fixed cards. If a hand has one around and two runs, it fixes 12 cards. There are $\binom{4}{2}\binom{4}{1} = 24$ ways to have two arounds and one run, and $\binom{4}{1}\binom{4}{2} = 24$ ways to have one around and two runs. Thus, $S_2 = 24\,\texttt{fix}\,(11) + 24\,\texttt{fix}\,(12)$. $S_3 = 0$ because all hands with a total of four (arounds plus runs) require more than 12 cards.

17. Because the run suit and marriage suit must be different, there are 12 ways to have one run and one marriage. Each combination fixes seven cards, so $S_1 = 12\,\texttt{fix}\,(7)$. There are $\binom{4}{2}\cdot 2 = 12$ ways to have two runs and a marriage (in a suit different from both run suits), and each way fixes twelve cards. Likewise, there are 12 ways to have one run and two marriages, and each of these hands fixes nine cards. Thus, $S_2 = 12\,\texttt{fix}\,(9) + 12\,\texttt{fix}\,(12)$. Three runs and one marriage, or two runs and two marriages, takes more than 12 cards; hence, the only nonzero contribution to S_3 is to have one run and all three non-trump marriages. There are 4 ways for this to occur, and each way fixes 11 cards. Thus, $S_3 = 4\,\texttt{fix}\,(11)$. Because $S_4 = 0$, the answer is $S_1 - S_2 + S_3 = 12f_7 - 12f_9 + 4f_{11} - 12f_{12}$.

18. $5\,\texttt{fix}\,(2) - \texttt{fix}\,(3) - 5\,\texttt{fix}\,(4) - 7\,\texttt{fix}\,(5) + 15\,\texttt{fix}\,(6) - 13\,\texttt{fix}\,(7) +$
$12\,\texttt{fix}\,(8) + 8\,\texttt{fix}\,(9) - 27\,\texttt{fix}\,(10) + 19\,\texttt{fix}\,(11) - 7\,\texttt{fix}\,(12)$.

19. $\texttt{fix}\,(6)/\,\texttt{fix}\,(0) \approx .00304813$, while distinct hands gives

$$\frac{330{,}145}{287{,}134{,}346} \approx .0011498.$$

For pinochles, we have $\texttt{fix}\,(2)/\,\texttt{fix}\,(0) \approx .1857$, while distinct counts gives (using **Fact 4**)

$$\frac{\sum_{i=0}^{2}\binom{2}{i}\left[\overline{\begin{smallmatrix}22\\10-i\end{smallmatrix}}\right]}{\left[\overline{\begin{smallmatrix}24\\12\end{smallmatrix}}\right]} = \frac{\sum_{i=0}^{2}\binom{22}{i}\binom{24-i}{10-2i}}{287{,}134{,}346} \approx .1044.$$

References

Comtet, L. 1974. *Advanced Combinatorics.* Dordrecht, Holland: Reidel.

Graham, Ronald L., Donald E. Knuth, and Oren Patashnik. 1994. *Concrete Mathematics.* 2nd ed. Reading, MA: Addison-Wesley.

Grimaldi, Ralph P. 1994. *Discrete and Combinatorial Mathematics: An Applied Introduction.* 3rd ed. Reading, MA: Addison-Wesley.

Gudder, Stanley P. 1980. Four of a kind in pinochle. *American Mathematical Monthly* 87: 297–299.

Tucker, Alan. 1989. *Applied Combinatorics.* 2nd ed. New York: Wiley.

About the Authors

Tom Linton is an assistant professor of mathematics at Western Oregon State College, in Monmouth, Oregon. His research interests are the foundations of analysis, applications of descriptive-set-theoretic games to various areas of mathematics, and applications to undergraduate mathematics that involve some form of computational technology. Tom received a B.A. degree in mathematics from St. Olaf College in 1983 and M.A. and Ph.D. degrees in mathematical logic from the University of Wisconsin–Madison. He spent two years at Caltech as a Bateman Postdoctoral Instructor. He spent 1993–94 as a sabbatical replacement at Willamette University, where the idea for this paper was sparked. Since early childhood, Tom has enjoyed mathematics and the outdoors, especially golf, fishing, and ice hockey. Tom regularly plays pinochle with in-laws from south central Montana!

Richard Iltis is a professor of mathematics at Willamette University in Salem, OR. His interests include calculus reform and probability models. He received a B.S. in engineering from South Dakota School of Mines and Technology and a Ph.D. in mathematics from the University of Oregon. He spent a year as a postdoctoral fellow at the University of Toronto. Richard does not play pinochle but has gone fishing with Tom.

UMAP

Modules in Undergraduate Mathematics and Its Applications

Published in cooperation with the Society for Industrial and Applied Mathematics, the Mathematical Association of America, the National Council of Teachers of Mathematics, the American Mathematical Association of Two-Year Colleges, The Institute of Management Sciences, and the American Statistical Association.

Module 746

How to Win at Nim

Daniel E. Loeb

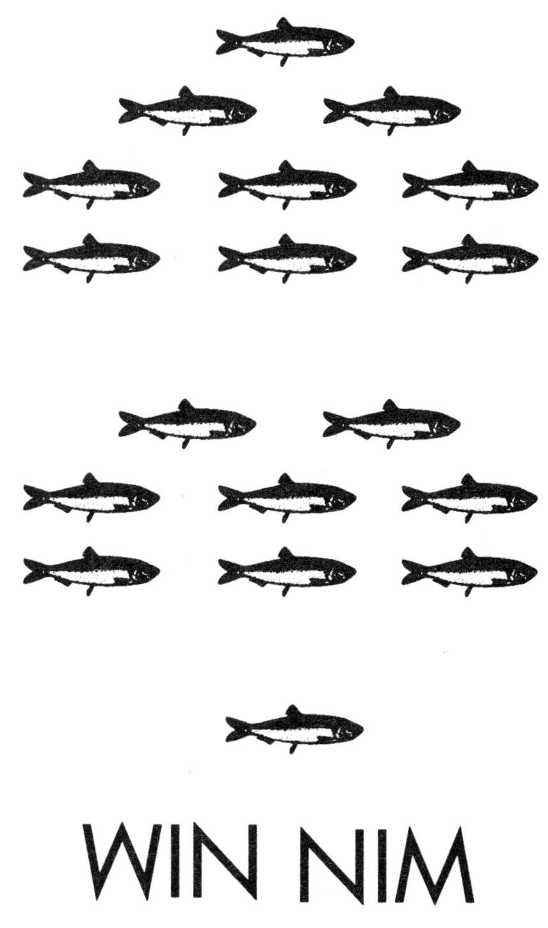

WIN NIM

Applications of Discrete Mathematics to Combinatorial Games

COMAP, Inc., Suite 210, 57 Bedford Street, Lexington, MA 02173 (617) 862–7878

INTERMODULAR DESCRIPTION SHEET:	UMAP Unit 746
TITLE:	How to Win at Nim

AUTHOR:	Daniel E. Loeb
	LaBRI
	Université de Bordeaux I
	33405 Talence Cedex
	France
	`loeb@labri.u-bordeaux.fr`
	`http://www.labri.u-bordeaux.fr/~loeb`

MATHEMATICAL FIELD:	Discrete mathematics, combinatorial game theory
APPLICATION FIELD:	Combinatorial games
TARGET AUDIENCE:	Undergraduate or high-school mathematics students in a course in combinatorics or discrete mathematics that places an emphasis on creativity and independent study.
ABSTRACT:	Nim is a very good example of an impartial game, since in some sense any impartial game can be "translated" into Nim. This module uses a variant of the Moore teaching method to stimulate the students to derive the theory of impartial games for themselves from the ground up. An appendix gives suggestions to the instructor.
PREREQUISITES:	Some knowledge of mathematical induction, sets, and binary arithmetic, together with total ignorance of the winning strategy for Nim and other impartial games.

COMAP, Inc., Suite 210, 57 Bedford Street, Lexington, MA 02173
(800) 77–COMAP = (800) 772–6627 (617) 862–7878

How to Win at Nim

Daniel E. Loeb
LaBRI
Université de Bordeaux I
33405 Talence Cedex
France
email: loeb@labri.u-bordeaux.fr
WWW: http://www.labri.u-bordeaux.fr/~loeb

Table of Contents

Modules and Monographs in Undergraduate Mathematics and its Applications (UMAP) Project

The goal of UMAP is to develop, through a community of users and developers, a system of instructional modules in undergraduate mathematics and its applications, to be used to supplement existing courses and from which complete courses may eventually be built.

The Project was guided by a National Advisory Board of mathematicians, scientists, and educators. UMAP was funded by a grant from the National Science Foundation and is now supported by the Consortium for Mathematics and Its Applications (COMAP), Inc., a nonprofit corporation engaged in research and development in mathematics education.

Paul J. Campbell	Editor
Solomon Garfunkel	Executive Director, COMAP

1. The Rules of Nim

Let us play *Nim*. All we need is two players and some matches (or coins or pieces of chalk). We arrange the matches arranged into several piles on an overhead projector. The players take turns choosing a pile and removing matches from it. At least one match, and up to an entire pile, can be removed in a single turn.

A player who cannot move loses.

Nim has been played by the Chinese (under the name *Fan Tan*) since antiquity.

How can this come about?

Only when there are no matches left. In other words, whoever takes the last match wins.

Exercise

1. Play the game for a while with a friend. You will quickly conjecture some rules that form the basis for the strategy of the game. Don't read on before trying this exercise, or you'll spoil the fun for yourself!

Here are a few pieces of advice that you may have discovered:

- When there is only one pile, take all of it and win.
- When there are two piles, equalize them. Your opponent will be obliged to disequalize them. Pursuing equalization to the exhaustion of the two piles, you win.
- On the other hand, when there are three unequal piles, avoid making them equal. Otherwise, your opponent will remove the third.

Exercise

2. Play some more Nim and try to formulate further strategies. Make lists of winning and losing positions. Which are there "more" of? Why?

2. Definition of "Winning"

Obviously, some positions are winning and others losing. But which is which? And what exactly does one have to do to check if a position is winning or losing? To be sure of what we are talking about, we need a rigorous definition of two terms that you probably already thought you understood: "winning" and "losing."

Let P be a position in a game G. Suppose that P leads to a choice of several positions $P_1, P_2, \ldots, P_n$. We say that P is *winning* (for the player whose turn it is) if there is at least one choice P_i that is losing. A *winning move* is then a choice of a losing position for the opponent. On the other hand, P is *losing* if all of the choices P_i are winning, so that a winning move is not possible.

Isn't this definition circular? In fact, "winning" is defined in terms of "losing," and vice versa! Fortunately, it *isn't* circular, because the games P_i are shorter than P. Formally, we have the following theorem.

Theorem. *Every position is either winning or losing, but not both.*

Proof: The proof is by induction on the number of moves that the game might last.

Let P be a position with no available moves. Hence, P is not winning, because there is no choice P_i that is losing (since there is no choice at all). On the other hand, all of the (nonexistent) choices P_i are winning, so P is losing.

Now, suppose that the theorem is true for games that last no more than $m-1$ moves, and let P be a position with up to m moves left. All of its options P_i have at most $m-1$ moves left. By hypothesis, each of them is winning or losing but not both. Hence, P itself is winning or losing but not both. □

Exercises

3. Prove that in the game of chess, one of the following holds true: White can force a win, Black can force a win, or both players can force at least a draw. (For extra credit, figure out which one it is!)

4. What do "winning" and "losing" mean in an "infinite" game, one with no upper bound on the number of moves?

5. What do "winning" and "losing" mean in a game with three players?

3. Nim-Sum

Let's concentrate now on Nim with three piles of matches. Let xyz denote one pile each of x, y, and z matches.

As you may have noticed in **Exercise 2**, most positions xyz are winning, so it is easier to learn the losing positions and try to trap the opponent into one.

The condition that xyz be losing is really quite strict; for xyz to be winning, it suffices to find a single losing option xyz', $xy'z$, or $x'yz$.

By definition, 000 (no matches) is losing, and we have already seen that $0xx$ (two equal piles) is losing for any x. The game 123 is losing because all of the choices are winning for the opponent:

If we leave...	023,	the opponent will leave...	022;
	113,		110;
	103,		101;
	122,		022;
	121,		101;
	120,		110.

Exercises

6. Show that the positions 145, 167, and 246 are losing. What other positions are losing?

7. Is it possible that both xyz and xyt are losing, if $z \neq t$? Don't read on before trying this exercise, or you'll spoil the fun for yourself!

The answer to **Exercise 7** is no. Suppose, without loss of generality, that $z > t$. Then xyt is a losing option from the position xyz, contradicting the idea that they may both be losing.

This property allows us to make the definition:

$$x \oplus y = \begin{cases} z, & \text{the unique number for which } xyz \text{ loses;} \\ \text{a red herring,} & \text{if there is no such number.} \end{cases}$$

There is definitely something fishy about this definition! But we can begin to fill out the table of values of $x \oplus y$, using the losing positions that we already know (see **Table 1**).

Table 1.
Values of $x \oplus y$.

$\oplus$	0	1	2	3	4	5	6	7	8	9
0	0	1	2	3	4	5	6	7	8	9
1	1	0	**3**	**2**	5	4	7	6		
2	2	**3**	0	**1**	6		4			
3	3	**2**	**1**	0						
4	4	5	6		0	1	2			
5	5	4			1	0				
6	6	7	4		2		0	1		
7	7	6					1	0		
8	8								0	
9	9									0

The fact that 123 is losing lets us write

$$1 \oplus 2 = 3, \quad 1 \oplus 3 = 2, \quad 2 \oplus 3 = 1, \quad 2 \oplus 1 = 3, \quad 3 \oplus 1 = 2, \quad \text{and} \quad 3 \oplus 2 = 1.$$

Thus, each new losing position lets us make up to six new entries in the table. The gaps in the table allow us to focus our attention on positions worthy of further study. Once completed, this table will contain all that is necessary for us to play Nim well with three piles.

Exercise

8. Fill out the rest of the entries in **Table 1**. (Write a computer program if you are so inclined.) Are you forced to use the red herring?

4. The "Mex" Rule

The table of $x \oplus y$ is very useful for being able to win at Nim. It allows us to replace unwieldy lists of losing positions with a convenient table. Nevertheless, the table is difficult to fill it out, especially if you begin to mix up "lose" and "win." The following rule gives us a way to calculate an entry $x \oplus y$ in the table from earlier values.

Theorem. *$x \oplus y$ equals the smallest number that is neither in the column above nor in the row to the left of its position. Hence, $x \oplus y$ is never a red herring.*

> **Example.** $4 \oplus 6 = 2$. The numbers 4, 5, 6, 7, 0, and 1 occur to the left of the position for $4 \oplus 6$, and 6, 7, 4, and 5 occur above that position; 2 is the smallest number not included in the set $\{0, 1, 4, 5, 6, 7\}$ (see **Table 2**).

Table 2.
Determining $4 \oplus 6$.

$x \oplus y$							6
							6
							7
							4
							5
4	4	5	6	7	0	1	?

The rule of the theorem is called the *"Mex"* (**minimum excluded** value) rule. In fact, $x \oplus y$ is the minimal number excluded from the set

$$S = \{x' \oplus y : x' < x\} \cup \{x \oplus y' : y' < y\}.$$

Proof: Let z be the value defined this way, as the "Mex" of x and y. We must show that xyz is losing. To show that, it suffices to show that all of the positions that xyz can lead to are winning.

Starting from xyz:

- If we take from the first pile, we arrive at a position $x'yz$ with $x' < x$. But $x' \oplus y$ is above $x \oplus y$ in the table. Hence, $z \neq x' \oplus y$, and so $x'yz$ is a winning position.
- If we take from the second pile, we arrive at a position $xy'z$ with $y' < y$. But $x \oplus y'$ is to the left of $x \oplus y$ in the table. Hence, $z \neq x \oplus y'$, and so $xy'z$ is a winning position.
- If we take from the third pile, we arrive at a position xyz' with $z' < z$. But z is the smallest number that is not to the left nor above $x \oplus y$. Hence, z' must be among the numbers S to the left or above $x \oplus y$. Suppose it is among the ones to the left. That means that there exists an $x' < x$ with $z' = x' \oplus y$. Then $x'yz'$ is losing. But $x'yz'$ is a position reachable directly from xyz', hence xyz' is a winning position. The other case is similar, and we leave it as an exercise. □

Exercises

9. Complete the proof.

10. The "Mex" rule allows us easily to enlarge the table of $x \oplus y$. To calculate $x \oplus y$, we need to calculate all of the numbers to the left and above $x \oplus y$ and take their "Mex." . We can then play Nim with three piles perfectly! Make a 17×17 table of $x \oplus y$ by hand, or program a computer to generate a 289×289 table. What do you discover?

11. Is $\oplus$ a "group" operation? That is:

- Is $\oplus$ associative? ($x \oplus (y \oplus z) = (x \oplus y) \oplus z$ for all x, y, z.)
- Is there an identity element e? ($e \oplus x = x = x \oplus e$ for all x.)
- Does every x have an "inverse" y? ($x \oplus y = e = y \oplus x$.)

5. Exclusive-Or

You probably noticed that the parts of the table of $x \oplus y$ repeat. The table consists of two-by-two blocks

a	b
b	a

arranged into four-by-four blocks

$$\begin{array}{|cc|cc|} a & b & c & d \\ b & a & d & c \\ \hline c & d & a & b \\ d & c & b & a \end{array},$$

and so forth.

This makes us think of powers of two. In fact, it is highly useful to rewrite the table *in base* 2.

Exercises

12. Fill out the following $\oplus$ table in binary:

$x \oplus y$	0000	0001	0010	0011	0100	0101	0110	0111	1000	1001
0000										
0001										
0011										
0100										
0101										
0110										
0111										
1000										
1001										

13. Does this table give you an idea about how to calculate $x \oplus y$ directly? Don't read on before trying this exercise, or you'll spoil the fun for yourself!

Looking over the table in binary, we see a new way to calculate $x \oplus y$. The table shows a 0 bit as place value in $x \oplus y$ in every position where x and y have the same bit, and a 1 where the bits of x and y are different. Computer scientists call this way of combining x and y *exclusive-or*. It's addition in binary for which we discard carries.

Example. Let a game of Nim have three piles with 170, 289, and 1993 matches. You don't want to have to write out a table of $x \oplus y$ that far! You simply calculate 170 $\oplus$ 289 and learn directly how many matches to take from the third pile.

In binary, 170 and 289 are represented as 10101010 and 100100001. Their Nim-sum is 110001011, or 395 in decimal. So, reduce the third pile to 395 matches.

$$\begin{array}{rcr} 170 & = & 10101010 \\ 289 & = & 100100001 \\ \hline 395 & = & 110001011 \end{array}$$

Exercises

14. Prove that this method of calculating $x \oplus y$ actually works as advertised. Generalize.

15. Is there a Nim-multiplication $\otimes$ that is "compatible" with $\oplus$?

6. Nim with Many Piles

Now that the game with three piles is resolved, the game with four or more piles is very simple. The general strategy is as follows:

Theorem. *Given a position in Nim, write the sizes of the piles in binary. Add them without carrying (i.e., use exclusive-or). If the result is 0, the position is losing; otherwise, the position is winning.*

Example. Given piles of sizes 1, 5, 7, 17, 23, and 34, their sum is 100111.

$$
\begin{array}{rcr}
1 & = & 1 \\
5 & = & 101 \\
7 & = & 111 \\
17 & = & 10001 \\
23 & = & 10111 \\
34 & = & 100010 \\
\hline
39 & = & 100111
\end{array}
\quad \longrightarrow \quad
\begin{array}{rcr}
1 & = & 1 \\
5 & = & 101 \\
7 & = & 111 \\
17 & = & 10001 \\
23 & = & 10111 \\
5 & = & \mathbf{000101} \\
\hline
0 & = & \mathbf{000000}
\end{array}
$$

So the position is winning. If you reduce the pile of 34 to 5 matches, the sum will be 0 and the opponent will have a losing position.

Proof: We prove the theorem by induction on the total number of matches.

If the total has a Nim sum of zero, we must show that any move will make the total nonzero. However, any move will affect at most one row of the "addition." In this addition, at least one bit will be changed. Since there were formerly an even number of 1s in that column there most now be an odd number of 1s, so at least one bit of the total is non-zero.

On the other hand, if the total has a nonzero Nim sum, then we must find a "winning move" that make the total zero again. Find any pile that contributes to the leftmost bit of the total. Reduce this pile to equal the Nim sum of the other piles by writing 0 in place of 1 and 1 in place of 0 for every bit in a column occupied by a 1 in the total (such bits are shown in **bold** in the example above.). Verify that this new pile is really smaller than the original pile. □

7. Conclusion

We are always sure to win at Nim, unless our opponent also knows the theorem above and by chance we start off in one of the rare losing positions. Alas, no one is going to want to play Nim with us!

So can we win any other games?

The Sprague-Grundy theorem [Sprague 1936; Grundy 1939] says yes.

As we have seen, Nim is played with a number of piles that do not interact. Similarly, many games can be thought of as breaking down into a number of "components," with each move affecting exactly one component. Players are free to play on the component of their choice. If there is no move available in any of the components, then the player loses.

Actually, each component is really a game of its own, so we can speak of playing a *sum* of games. The game $G + H$ is the game in which players take turns making a move in G or in H (but not both at the same time), and the game ends when there is no legal move left in either game. The player who was supposed to move is considered to have lost.

Let G and H be *impartial* games. This means that they are games (like Nim) in which the moves available to either player in any given position are exactly the same. (For example, chess is not impartial since only White is allowed to move the White pieces.) The games G and H are considered *equivalent* (written $G \sim H$) if you can replace one with the other in a sum without affecting the outcome of the game. In other words, $G \sim H$ if and only if given any game K,the game $G + K$ is winning exactly when $H + K$ is winning.

Exercise

16. Show that *equivalent* is an equivalence relation.

17. Show that all losing positions in any game are equivalent to an empty Nim pile.

18. Although any nonempty Nim pile is a winning position, show that no two Nim piles are equivalent unless they are equal in size.

The Sprague-Grundy theorem says that *up to equivalence* Nim is the only impartial game. That is, all impartial games are *really* Nim in disguise. In other words, for every impartial game G, one can create a "dictionary" that translates each position P into an equivalent position $\mathcal{G}(P)$ in a game of Nim with one pile. One can indeed calculate $\mathcal{G}(P)$. It is the smallest number not equal to $\mathcal{G}(Q)$ for every option Q from the position P. We leave detailed consideration of the Sprague-Grundy theorem to a projected sequel Module.

Since you know how to play Nim well, and since all impartial games are equivalent to Nim with one pile, you know how to play all impartial games well!

Exercises

19. Explore the following games (or your own impartial games):

a) *Queens.* A number of queens are placed on a checkerboard. Each queen can move any number of squares north, west, or northwest. Each player moves exactly one queen each move. Eventually, all the queens will reach the northwest corner, and the player who makes the last move wins. (Hint: How does it change the strategy if we suppose that two queens kill each other if one lands on the square of the other?)

b) *Wythoff's Game.* Consider a variant of the above using knights instead of queens. (Why isn't it interesting to study the obvious variant using rooks?)

c) *Limited Nim.* This is just like Nim except that you may not remove more than k matches from a pile in a single move.

d) *How to Cut a Cake Well.* Mom has prepared a rectangular cake, n inches by m inches. Alice and Bill in turn cut pieces in two until there remain nm pieces, each 1 inch by 1 inch. The last person to cut a piece eats the whole cake; the other does the washing up. (Hint: This game is not as hard as it looks.)

e) *Impartial Domineering.* The players in turn place dominos on a checkerboard. The last player to place a domino wins. (Hint: Consider first $1 \times n$ checkerboards.)

f) *Grundy's Game.* We play with piles of matches, and the only move allowed is to divide a pile into two unequal piles. The game ends when all of the piles contain one or two matches. The player who plays last wins. (Open question: Is $\mathcal{G}(n)$ eventually periodic with period three?)

g) *Checkers.* A white checker and a black checker are placed on different squares of each row of a checkerboard of arbitrary size. A legal moves is for white player to slide any one of his checkers any number of squares right or left while remaining on the same row. But there's a catch, white can't jump over black (or land on black). Black moves similarly. Someone loses if they can't move (in other words, all their checkers are pinned against one wall or the other.)

h) *Bowling.* We are expert bowlers trying to knock down a row of n pins. We are capable of knocking down any one pin or two adjacent pins, but the pins are too far apart to hope to knock down three or more pins. Whoever knocks down the last pin wins the game.

i) *Tic-Tac-Toe.* In this poor man's version of Tic-Tac-Toe, both players take turns making the same sign "×" on a $1 \times n$ board. The winner is the first player to complete a sequence of three or more "×"s.

j) *Loser's Nim.* This is just like Nim except that the player who removes the last match is the loser instead of the winner.

k) *Dim.* In the game of Dim, you may remove m matches from a pile of n matches provided that n is a multiple of m. You may completely remove a pile of n matches provided that $n \geq 2$. (Hint: Prime numbers are important in the strategy of this game.)

8. Solutions to Exercises

1. See p. 1.

2. See **Section 3**.

3. The rules of chess provide that the game is drawn if a player demonstrates that 50 moves have been made by each player without a capture or a pawn move. Thus, no game of chess can last more than 12,700 moves. (Even without this rule, a larger upper bound is imposed by the rule that if the same position occurs three times with the same player to move, that player may claim a draw.) In other words, chess is a finite game. Hence, we can apply the theorem of **Section 2** if we award the draws to Black. Thus, either White can win or Black can force a win or a draw. Similarly, Black can win or White can force a win or a draw.

4. Some positions (called draws) are neither wins nor losses. The Sprague-Grundy theory mentioned in **Section 7** can be generalized to this context. However, $\mathcal{G}(P)$ takes various "infinite" values. See Berlekamp et al. [1982] for details.

5. Several generalizations are possible; for example, see Propp [1996] or Loeb [1996].

6.

145		167		246	
If we leave...	the opponent will leave...	If we leave...	the opponent will leave...	If we leave...	the opponent will leave...
045	044	067	066	146	145
135	132	157	154	046	044
125	123	147	145	236	231
115	110	137	132	226	220
105	101	127	123	216	213
144	044	117	110	206	202
143	123	107	101	245	145
142	132	166	066	244	044
141	101	165	145	243	213
140	110	164	154	242	213
		163	123	241	231
		162	132	240	220
		161	101		
		160	110		

7. See p. 3.

Table 3.
The table for the operation $\oplus$.

$\oplus$	0	1	2	3	4	5	6	7	8	9
0	0	1	2	3	4	5	6	7	8	9
1	1	0	3	2	5	4	7	6	9	8
2	2	3	0	1	6	7	4	5	10	11
3	3	2	1	0	7	6	5	4	11	10
4	4	5	6	7	0	1	2	3	12	13
5	5	4	7	6	1	0	3	2	13	12
6	6	7	4	5	2	3	0	1	14	15
7	7	6	5	4	3	2	1	0	15	14
8	8	9	10	11	12	13	14	15	0	1
9	9	8	11	10	13	12	15	14	1	0

8. No red herring is needed. See **Table 3**.

9. Otherwise, z' must be among the numbers S to "above" $x \oplus y$. That means that there exists a $y' < y$ with $z' = x \oplus y'$. Then $xy'z'$ is losing. But $xy'z'$ is a position reachable directly from xyz', hence xyz' is a losing position.

10. See **Section 5** for help.

11. Yes, it is a group operation. Since $0xx$ is a losing position, we have $0 \oplus x = x = x \oplus 0$ and $x \oplus x = 0$. Thus, 0 is the identity element, and every number x is its own inverse! Associativity is less obvious but follows from **Exercise 13**.

12. See **Table 4**.

Table 4.
Table for the operation $\oplus$, in binary.

$x \oplus y$	0000	0001	0010	0011	0100	0101	0110	0111	1000	1001
0000	0000	0001	0010	0011	0100	0101	0110	0111	1000	1001
0001	0001	0000	0011	0010	0101	0100	0111	0110	1001	1000
0010	0010	0011	0000	0001	0110	0111	0100	0101	1010	1011
0011	0011	0010	0001	0000	0110	0111	0100	0101	1011	1010
0100	0100	0101	0110	0111	0000	0001	0010	0011	1100	1101
0101	0101	0100	0111	0110	0001	0000	0011	0010	1101	1100
0110	0110	0111	0100	0101	0010	0011	0000	0001	1110	1111
0111	0110	0111	0100	0101	0011	0010	0001	0000	1111	1110
1000	1000	1001	1010	1011	1100	1101	1110	1111	0000	0001
1001	1001	1000	1011	1010	1101	1100	1111	1110	0001	0000

13. See p. 6.

14. This is a special case of the theorem in **Section 6**.

15. Yes; see **Table 5**. The operation $x \otimes y$ arises in the game Turing Corners described in Guy [1989, 68–69]. The numbers 0 through $2^{2^k} - 1$ form a finite field of characteristic zero with the operations $\oplus$ and $\otimes$.

Table 5.

Table for the operation $\otimes$.

$\otimes$	0	1	2	3	4	5	6	7	8	9	10	11	12	13	14	15
0	0	0	0	0	0	0	0	0	0	0	0	0	0	0	0	0
1	0	1	2	3	4	5	6	7	8	9	10	11	12	13	14	15
2	0	2	3	1	8	10	11	9	12	14	15	13	4	6	7	5
3	0	3	1	2	12	15	13	14	4	7	5	6	8	11	9	10
4	0	4	8	12	6	2	14	10	11	15	3	7	13	9	5	1
5	0	5	10	15	2	7	8	13	3	6	9	12	1	4	11	14
6	0	6	11	13	14	8	5	3	7	1	12	10	9	15	2	4
7	0	7	9	14	10	13	3	4	15	8	6	1	5	2	12	11
8	0	8	12	4	11	3	7	15	13	5	1	9	6	14	10	2
9	0	9	14	7	15	6	1	8	6	12	11	2	10	3	4	13
10	0	10	15	5	3	9	12	6	1	11	14	4	2	8	13	7
11	0	11	13	6	7	12	10	1	9	2	4	15	14	5	3	8
12	0	12	4	8	13	1	9	5	6	10	2	14	11	7	15	3
13	0	13	6	11	9	4	15	2	14	3	8	5	7	10	1	12
14	0	14	7	9	5	11	2	12	10	4	13	3	15	1	8	6
15	0	15	5	10	1	14	4	11	2	13	7	8	3	12	6	9

16. *Reflexivity*: Clearly, $G + K$ is winning if and only if $G + K$ is winning!
Symmetry: $G \sim H$ and $H \sim G$ both mean that $G + K$ is winning if and only if $H + K$ is winning.
Transitivity: Suppose that $G_1 \sim G_2 \sim G_3$. Then ($G_1 + K$ winning) $\Leftrightarrow$ ($G_2 + K$ winning) $\Leftrightarrow$ ($G_3 + K$ winning), so by the transitivity of logical implication we have G_1 winning if and only if G_3 is winning.

17. First note that the empty pile gives no possibility of movement, so the rules of the game K are exactly the same as that of the game $0 + K$. Now, let G be a losing position. Suppose K is a losing position. When playing $G + K$, whatever move we make in G or K can be immediately countered in *that* game. Thus, $G + K$ is a losing position. On the other hand, if K is a winning position, then there is a winning move K'. When playing $G + K$, we play K' in the second component, and win by the above reasoning.

18. Let G and K be piles of n matches and H a pile of m matches, with $n \neq m$. Clearly, $G + K$ is losing, but $H + K$ is winning. Thus, G and H are not equivalent.

19. Most of these games are mentioned in Berlekamp et al. [1982]. All of the games except Checkers and Loser's Nim are impartial and the Sprague-Grundy theorem can be directly applied. The student should find the elementary components P into which each game breaks down, for example, a single queen in the game Queens, or an isolated empty part of the board in Domineering or Tic-Tac-Toe. In some cases, a certain amount of thought will be necessary to convince yourself that the components are indeed *independent*. Each can be thought of as a Nim pile of size $\mathcal{G}(P)$.

Checkers in part **g** is a disguised version of Nim. Until the players realize that, the game is unlikely to end!

How to Cut a Cake Well is actually not much of a game at all. No matter how well or badly one plays, the winner is always the same!

Appendix: Use in the Classroom

The objective of this module is to introduce the students to creative mathematics. Mathematics is not a spectator sport, so we must ask the student to participate in its conception. The subject of impartial games was chosen for several reasons:

- All students have some experience playing games and can apply their game-playing intuition to the mathematical problems at hand.
- The students are motivated by their work (knowing that this work will lead to victory against students not in the class).
- The material has virtually no formal prerequisites yet is closely related to exciting on-going research.

Of course, the process of discovery is only relevant if all of the students share in it. Thus, before starting this Module, students should be asked if they have ever heard of the game of Nim. (If so, they should be given alternate material to work on and instructions not to discuss the topic with the others until the others have "caught up.") Similarly, while this material is both accessible and challenging to a wide range of students from junior high school to graduate school, it is recommended that the level of mathematical maturity be kept as homogeneous as possible with each workgroup.

Moreover, everyone should be given a chance to participate. Hence, classes should be kept small (under say 20). The teacher should act as a moderator and make a particular effort to draw everyone into the discussion. A talkative student can be appointed as scribe each day and record the progress made, in terms of definitions given, and conjectures proposed, refuted, or proved.

"Freeing" the class to discover on its own actually is more work for the teacher than the usual sort of class. The method proposed here is less structured than, for example, the Moore method (see [Chalice 1995]). Students will not necessarily discover everything in the order given in the Module, nor will they employ the same notation in their definitions as we have here. They will surely make unanticipated conjectures. These conjectures should not be dismissed out of hand, but they may be deferred to the next day or to a problem set if they are not convenient. The teacher must keep track of class discussion and adapt plans accordingly each evening.

The introduction to the Module must begin with a clear explanation of the rules of Nim. The teacher should play a single game against a student "volunteer" in order to make sure everyone is thinking of the same game. The teacher should obviously avoid giving away the optimal strategy in playing this sample game.

Some elements of the strategy are fairly simple, and each student should be given an opportunity to discover them. Thus, without any further discussion of the game, the class should break into groups of two or three and play the game.

The definition of "winning" might seem a little abstract for many students. However, it can be motivated in an ensuing class discussion, for each group will surely have invented its own notation. Some will have abandoned matches and begun working with numbers written on paper. Some will call two equal piles "winning" and others will call two equal piles "losing." A rose by any other name might smell as sweet, but the class will have to come to grips with the differences in nomenclature and chose *something* standard as a common ground for future discussion. It will be especially helpful if one group happens to disagree with another group about whether a certain position is winning. This type of argument can only be settled if the students are clear on what needs to be proven to show that something is winning, and this can only come from a well-written definition.

The Module is probably best done over a period of many days while other material is studied in parallel. Subtle hints by the teacher may need to given if the entire class becomes blocked in their work for a day or two.

The solution to most of the exercises appears later in the text and should not be revealed immediately to the students. Other exercises are open-ended and are treated partially in the references. These references can also serve as a good source of material if the class continues to show interest in combinatorial game theory after the completion of this module.

To emphasize the important role of Nim in the context of the theory of impartial games, each student could choose or be assigned a game to work on at the end of the Module. Based on Sprague and Grundy's theorem, the student will be able to translate the particular game into the terms of Nim and be be able to simplify work on the game by making use of the class's work on Nim. Games should be carefully chosen according to the preferences and capacities of each student. The assignment of a new game could be considered an independent study project, and the students could be required to write an essay or give an oral presentation explaining their game, its optimal strategy, and any interesting properties.

Acknowledgments

The material presented here formed part of the basis of a workshop "Surreal Numbers and Games" at three summer programs for young scholars: Université Mathématique d'Été (ENSAE, Toulouse, France, 1991), and the Hampshire College Summer Studies in Mathematics (Hampshire College, Amherst, MA, USA, 1993 and 1994). I thank my students and my junior staff (Paul Karafiol, Jeremy Weinstein, Cathy O'Neil, Mark Nowakowski, Josh Mandell, and Yihao Zhang) for their enthusiastic support.

Parts of this Module has also been tested in short afternoon workshop (for example, at the Fédération Française de Jeux Mathématiques's summer camp (Parthenay, France, 1992) and at the French national youth congress sponsored by MATh.en.JEANS (École Polytechnique, Palaiseau, France, 1993, and Université de Cergy, France, 1994). MATh.en.JEANS is a non-profit organization that sponsors math clubs at high schools and junior high schools throughout France. Students work on mini-research projects proposed by local mathematicians. This Module appeared originally as "Gagner au Jeu de Nim" in the proceedings of the 1993 MATh.en.JEANS conference.

The author was partially supported by URA CNRS 1304, NATO CRG 930554, EC Grant CHRX–CT93–0400, and PRC Math-Info.

References

Beck, Anatole. 1969. Games. Chapter 5 in *Excursions into Mathematics*, by Anatole Beck, Michael N. Bleicher, and Donald W. Crowe, pp. 315–387. New York: Worth Publishers.

Introduces the notions of impartial games and winning/losing positions with numerous examples and exercises.

Berlekamp, Elwyn R., John H. Conway, and Richard K. Guy. 1982. *Winning Ways for Your Mathematical Plays.* Vol. 1: *Games in General.* Vol. 2: *Games in Particular.* New York: Academic Press.

This is the premier book about combinatorial game theory in general and about many of your favorite recreational games. The appearance of this book set off the current flurry of research on combinatorial games.

Bouton, Charles L. 1901. Nim, a game with a complete mathematical theory. *Annals of Mathematics* 2nd ser. 3(1) (October 1901) 35–39.

Bouton completely solved Nim with an arbitrary number of piles of arbitrary size. To see why he called the game NIM, just turn this page 180 degrees!

Chalice, Donald R. 1995. How to teach a class by the modified Moore method. *American Mathematical Monthly* 102: 317–321.

Conway, J.H. 1976. *On Numbers and Games.* New York: Academic.

This book is the precursor to Berlekamp et al. [1982] (*"Winning Ways"*). Conway shows how certain partizan games can be thought of as "surreal" numbers (finite, real, or even infinitesimal).

Fraenkel, Aviezri. 1994. Dynamic bibliography on combinatorial games. *Electronic Journal of Combinatorics, Dynamic Surveys*, DS2.

A comprehensive survey of the current state of the art in the theory of combinatorial games. Over 500 references; continually updated by the author.

Gleason, Andrew. 1966. *Nim and Other Oriented-Graph Games.* 16 mm B&W film, 63 min. Mathematical Association of America. Distributed by Ward's Modern Learning Aids Division (P.O. Box 1712, Rochester, NY 14603 or P.O. Box 1749, Monterey, CA 93940).

This film has not been reissued by the Mathematical Association of America on videotape. It may still be available from the university film rental libraries at Idaho State University, Pennsylvania State University, and others.

Grundy, P.M. 1939. Mathematics and games. *Eureka* 2 (1939) 6–8.

Grundy (a student in statistics at Cambridge University) independently discovered how all impartial games could be translated into Nim.

Guy, Richard K. 1989. *Fair Game: How to Play Impartial Combinatorial Games.* Arlington, MA: COMAP.

Contains short chapters on several dozen games, together with exercises and student projects.

Knuth, Donald K. 1974. *Surreal Numbers.* Reading, MA: Addison Wesley.

This excellent book is written in the form of a play. It is the story of a girl and a boy who rediscover Conway's beautiful theory of surreal numbers. In the process (in particular, see sections 7 and 11), the characters discuss the true nature of mathematics and how it should be taught. These "surreal numbers" actually represent certain kinds of games. In fact, the *symmetric pseudo-numbers* mentioned at the end of the book correspond exactly to the game Nim.

Loeb, Daniel E. 1996. Stable winning coalitions. To appear in *Proceedings of the Workshop on Combinatorial Games at MSRI,* edited by S. Levy, R. Nowakowski, and R. Guy. New York: Cambridge University Press.

A survey of several different methods proposed for classifying multiplayer games.

O'Beirne, T.H. 1965. *Puzzles and Paradoxes.* New York: Oxford University Press. 1984. Reprinted under the title *Puzzles and Paradoxes: Fascinating Excursions in Recreational Mathematics.* New York: Dover.

Chapter 8, "Gamesman, Spare that Tree" (pp. 130–150), and Chapter 9, "'Nim' You Say, and 'Nim' It Is" (pp. 151–167), treat impartial games, including Nim, Wythoff's game, Grundy's game, and Welter's game.

Propp, James G. 1996. Three player impartial games. *Theoretical Computer Science.* To appear.

An analysis of forced wins in sums of three-player impartial games. Special attention is paid to three-player Nim.

Sprague, R. 1936. Über mathematische Kampfspiele. *Tohoku Mathematical Journal* 41 (1936) 438–444.

While Sprague was the original discoverer of the Sprague-Grundy theorem, his contribution went largely unnoticed for many years. Sprague published his paper in German in a Japanese journal. Because of wartime restrictions, few copies of the journal were printed, and it was not distributed to Allied countries.

Vajda, Steven. 1992. *Mathematical Games and How to Play Them.* Chichester, England: Ellis Horwood.

Chapter 2 (pp. 21–85) treats two-player impartial games.

Wolfe, David. 1994. Gamesman's Toolkit. Computer program in C with source. Extended by Michael Ernst to handle Ancient Konane and Modern Konane. Available by sending email to `wolfe@cs.berkeley.edu` or to `mernst@lcs.mit.edu`.

Analyzes values of positions for the games Domineering, Toads and Frogs, and Ancient and Modern Konane.

________. 1995. Undergraduate research opportunities in combinatorial games. *The UMAP Journal* 16 (1): 71–77.

A gentle introduction to the ideas of combinatorial game theory, using the game of partizan Domineering. Suggests where to find other games to explore.

About the Author

Daniel Elliott Loeb is *chargé de recherche* for the CNRS at the Université de Bordeaux in France and is a member of the editorial board of the Mathematical Games section of the *Journal of Theoretical Computer Science.* His Ph.D. thesis (MIT, 1989) was on a logarithmic generalization of Gian-Carlo Rota's umbral calculus. In addition to his research on artificial intelligence, extremal combinatorics, game theory, and umbral calculus, Loeb advocates mathematical education. He organized the first summer mathematics program in France of the Young Scholar kind (1990, 1991); he is vice-president of MATH pour TOUS, which among its other activities to popularize mathematics administers the French Kangaroo contest to 100,000 French high school and university students each year; and he is an officer of MATh.en.JEANS, which organizes math clubs at schools around France.

UMAP

Modules in Undergraduate Mathematics and Its Applications

Published in cooperation with the Society for Industrial and Applied Mathematics, the Mathematical Association of America, the National Council of Teachers of Mathematics, the American Mathematical Association of Two-Year Colleges, The Institute of Management Sciences, and the American Statistical Association.

Module 747

The Terminator and Other Geographical Curves

Yves Nievergelt

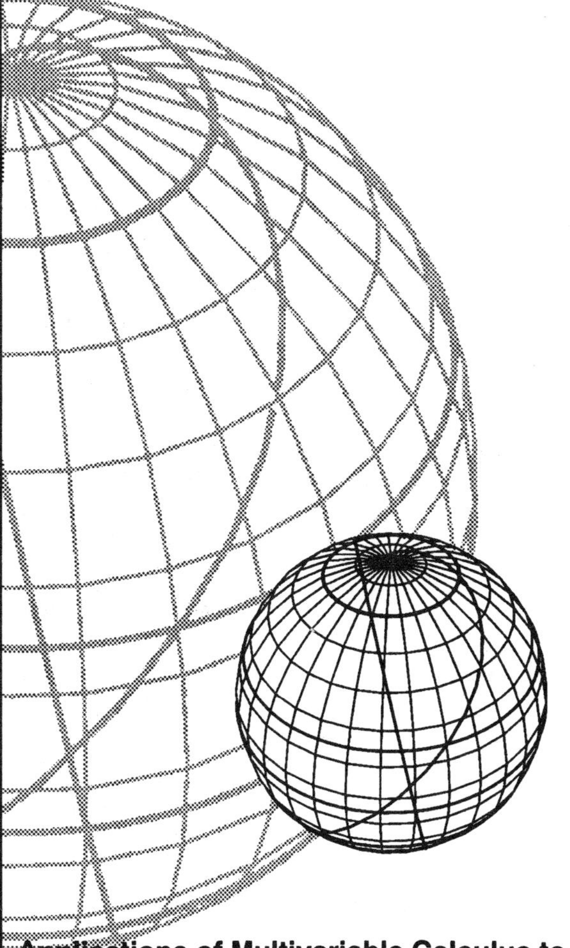

Applications of Multivariable Calculus to Cartography and Navigation

COMAP, Inc., Suite 210, 57 Bedford Street, Lexington, MA 02173 (617) 862–7878

INTERMODULAR DESCRIPTION SHEET: UMAP Unit 747

TITLE: The Terminator and Other Geographical Curves

AUTHOR: Yves Nievergelt
Dept. of Mathematics, MS 32
Eastern Washington University
526 5th Street
Cheney, WA 99004–2431
ynievergelt@ewu.edu

MATHEMATICAL FIELD: Introductory multivariable calculus

APPLICATION FIELD: Cartography, navigation

TARGET AUDIENCE: Students in multivariable calculus

ABSTRACT: This Module seeks to

- strengthen students' intuition in multidimensional geometry;
- consolidate their command of differentials;
- provide exercises, at a level between mechanical and theoretical, that do not require any background on applications; and
- show important applications that rarely appear in calculus texts.

We show how to use differentials of vector-valued functions to determine the angle between two routes, or the azimuth (direction) for the shortest route from one point to another, on the surface of the Earth. Real examples accompany the theory.

The exercises are at a level intermediate between mechanical problems and abstract proofs and fit in any multivariable calculus course.

PREREQUISITES: Prerequisite skills include the ability to ponder a problem that does not come with a canned recipe for its solution, and, from calculus: first derivatives of elementary functions; from linear algebra: matrices and compositions of linear transformations; from multivariable calculus: vector algebra, the concepts of curves in the plane and in space, surfaces, first derivatives of vector-valued functions (curves), and first and second partial derivatives.

RELATED UNITS:

Unit 748: *Differentials and Geographical Maps*, by Yves Nievergelt. Demonstrates loxodromes and the use of differentials of transformations of several variables to determine whether a geographical map preserves areas, preserves the magnitude and the orientation of angles, or neither.

Unit 206: *Mercator's World Map and the Calculus*, by Philip M. Tuchinsky. Reprinted in *UMAP Modules: Tools for Teaching 1977–79*, 677–727. Boston: Birkhäuser, 1980.

COMAP, Inc., Suite 210, 57 Bedford Street, Lexington, MA 02173
(800) 77–COMAP = (800) 772–6627 (617) 862–7878

The Terminator and Other Geographical Curves

Yves Nievergelt
Dept. of Mathematics, MS 32
Eastern Washington University
526 5th Street
Cheney, WA 99004–2431
`ynievergelt@ewu.edu`

Table of Contents

Modules and Monographs in Undergraduate Mathematics and its Applications (UMAP) Project

The goal of UMAP is to develop, through a community of users and developers, a system of instructional modules in undergraduate mathematics and its applications, to be used to supplement existing courses and from which complete courses may eventually be built.

The Project was guided by a National Advisory Board of mathematicians, scientists, and educators. UMAP was funded by a grant from the National Science Foundation and now is supported by the Consortium for Mathematics and Its Applications (COMAP), Inc., a nonprofit corporation engaged in research and development in mathematics education.

He had bought a large map representing the sea,
  Without the least vestige of land:
And the crew were much pleased when they found it to be
  A map they could all understand.

"What's the good of Mercator's North Poles and Equators,
  Tropics, Zones, and Meridian Lines?"
So the Bellman would cry: and the crew would reply,
  "They are merely conventional signs!"

"Other maps are such shapes, with their islands and capes!
  But we've got our brave Captain to thank"
(So the crew would protest) "that he's bought *us* the best —
  A perfect and absolute blank!"

Lewis Carroll, *The Hunting of the Snark. An Agony in Eight Fits.*
London, 1876.

1. Introduction

This Module and its sequel, *Differentials and Geographical Maps* (Module 748), demonstrate real applications, to cartography and to navigation, of the mathematical concept of the *differential* of a function of several variables. These applications include the design of geographical maps, called *conformal* or *isogonal* maps, that show angles as they appear on the surface of a spherical planet. Examples are stereographic projections (for polar expeditions) and Mercator projections (for navigators at sea), which let travelers measure their bearings on the chart.

The design of such maps requires differentials of functions from the plane of the map to the surface of the Earth in space. In preparation for that task, this Module focuses on mathematically simpler geographical *curves*. Indeed, great circles and other curves in space involve only the ordinary derivative of vector-valued functions, and level curves on a map involve only the first partial derivatives of real-valued functions. One of the applications presented here consists in calculating the direction of the shortest route joining two points on the surface of the Earth.

2. Review of Differentials

This section establishes notation while reviewing elementary differentials from a point of view that will help in the transition to differentials with several variables. We include only the theory for which most students need a review, and most of the examples and exercises pertain to the use of differentials in designing geographical maps.

2.1 Differentials with One Variable

Consider a function $f : \mathcal{D} \subseteq \mathbb{R} \to \mathbb{R}$ defined on an open set $\mathcal{D} \subseteq \mathbb{R}$ and with values on the real line $\mathbb{R}$. Recall from elementary calculus that the *derivative* $f'(x)$ of f at a point x is the limit of the slopes of lines passing through $(x, f(x))$ and another point $(x+h, f(x+h))$:

$$f'(x) = \lim_{h \to 0} \frac{f(x+h) - f(x)}{h}, \tag{1}$$

provided that the limit exists. The limit means that for every positive number ε, which measures the tolerance between $[f(x+h) - f(x)]/h$ and the value $f'(x)$ of the limit, a positive number δ exists, which measures the allowed variation h from x, such that

$$\left| \frac{f(x+h) - f(x)}{h} - f'(x) \right| < \varepsilon \tag{2}$$

for every h such that $0 < |h| < \delta$.

Application 1. If $f(t)$ represents the position of an object at time t, then $[f(t+h) - f(t)]/h$ represents the average velocity—the ratio of distance traveled over time—of that object in the interval of time from t to $t+h$, and the limit $f'(t)$ represents the velocity of that object at time t. The velocity takes into acount the direction of motion by means of the sign of $f'(t)$. Its absolute value, called the *speed*, records only the magnitude $|f'(t)|$. Similarly, $|[f(t+h) - f(t)]/h|$ represents the average speed. □

In different contexts, however, other definitions of the same derivative prove more useful [Thurston 1994]. For example, multiplying both sides of **(2)** by $|h|$ gives

$$|f(x+h) - \{f(x) + h\,f'(x)\}| < \varepsilon|h|. \tag{3}$$

This inequality describes how closely the expression $f(x) + h\,f'(x)$ approximates $f(x+h)$ as h tends to zero. Moreover, the expression $f(x) + h\,f'(x)$ corresponds to a straight line $T_{f,x}$ with the point-slope equation

$$T_{f,x}(x+h) = f(x) + h\,f'(x),$$

where $x+h$ stands for any real number, so that $(x+h, T_{f,x}(x+h))$ denotes any point on the line $T_{f,x}$.

For the purpose of extracting qualitative information, yet another interpretation of the derivative reveals that the line $T_{f,x}$, which passes through $(x, f(x))$ with slope $f'(x)$, coincides with the straight line that remains closest to f in some neighborhood of $(x, f(x))$. In other words, every other line L through $(x, f(x))$ with a slope m different from $f'(x)$ stays farther away from f than $T_{f,x}$ does at every point $x+h \neq x$ in some neighborhood of $(x, f(x))$, as follows.

Proposition 1. *There exists positive* δ *such that for each number* h *for which* $0 < |h| < \delta$, *the following inequality holds:*

$$|f(x+h) - L(x+h)| > |f(x+h) - T_{f,x}(x+h)|.$$

Proof: To verify the inequality, write the point-slope equation of L in the form $L(x+h) = f(x)+h{\cdot}m$, and compare $|f(x+h) - L(x+h)|$ to $|f(x+h) - T_{f,x}(x+h)|$ in the open interval $]x-\delta, x+\delta[$, with δ corresponding to any ε such that $0 < \varepsilon < |m - f'(x)|/2$. For such an ε, $|f(x+h) - T_{f,x}(x+h)| < \varepsilon < |m - f'(x)|/2$, whence the reverse triangle inequality and **(3)** give

$$\begin{aligned}
&|f(x+h) - L(x+h)| \\
&\quad = |f(x+h) - \{f(x) + h \cdot m\}| \\
&\quad = |f(x+h) - \{f(x) + h\,f'(x) - h\,f'(x) + h \cdot m\}| \\
&\quad = |f(x+h) - \{f(x) + h\,f'(x)\} - \{h \cdot m - h\,f'(x)\}| \\
&\quad = |f(x+h) - \{f(x) + h\,f'(x)\} - h \cdot \{m - f'(x)\}| \\
&\quad \geq |\,|f(x+h) - \{f(x) + h\,f'(x)\}| - |h \cdot \{m - f'(x)\}|\,| \\
&\quad = |\,|h \cdot \{m - f'(x)\}| - |f(x+h) - \{f(x) + h\,f'(x)\}|\,| \\
&\quad = |h \cdot \{m - f'(x)\}| - |f(x+h) - \{f(x) + h\,f'(x)\}| \\
&\quad \geq |h| \cdot |m - f'(x)| - \varepsilon|h| \\
&\quad > 2\varepsilon|h| - \varepsilon|h| = \varepsilon|h| \\
&\quad > |f(x+h) - T_{f,x}(x+h)|.
\end{aligned}$$

Therefore, $T_{f,x}$ remains closer to f than L does. □

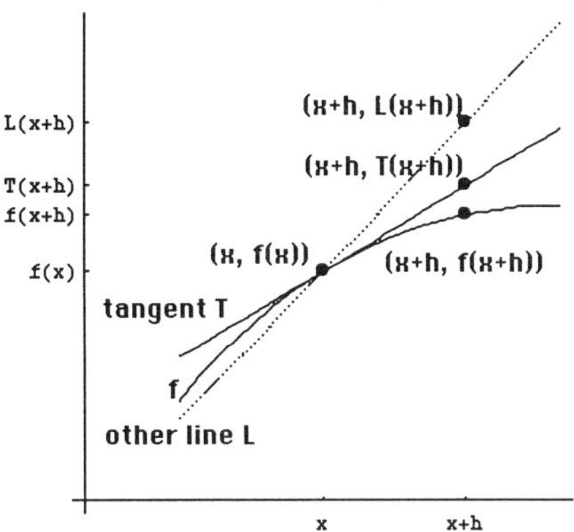

Figure 1. The tangent line $T = T_{f,x}$ remains closer to f near x than any other line L.

For this reason, $T_{f,x}$ is called the line *tangent* to f at $(x, f(x))$. Thus, *the derivative* $f'(x)$ *represents the slope of the tangent line* $T_{f,x}$. **Figure 1** illustrates the situation.

The foregoing considerations yield several interpretations of the derivative:

- the limit of slopes of lines through two points on the function,
- the linear part of the closest affine approximation of a function near one point,
- the slope of the line tangent to a function at one point.

Because the tangent line depends upon the location of the point of tangency, another related concept also proves useful: the *differential* of f at x is the function Df_x defined by the formula

$$Df_x(h) = h\, f'(x).$$

Thus, the differential is the linear part of the point-slope formula for the tangent line. As shown in the exercises, the differential remains invariant under translations and dilations. In particular, a picture of the differential remains invariant with or without a "microscope" [Thurston 1994].

The tangent line gives quantitative information about a function, whereas the differential gives qualitative information. For instance, the tangent line provides numerical (quantitative) approximations of the values of a function, and hence forms the basis for Newton-Raphson-Simpson method to solve nonlinear equations [Strang 1991, Ch. 3]. In contrast, the differential determines the location of the extrema (minima and maxima) and stationary points (qualitative information) for a function [Spivak 1980, Ch. 11; Strang 1991, Ch. 3]. The mean value theorem for derivatives and the corollary below demonstrate qualitative use of the derivative and differential.

Theorem 1. Mean Value Theorem for Derivatives. *If f is continuous at every point of the closed interval $[a, b]$, and if $f'(x)$ exists at every point of the open interval $]a, b[$, then a point $c \in]a, b[$ exists for which $f(b) - f(a) = (b-a)\, f'(c)$.*

Proof: See Spivak [1980, 179] or Strang [1991, §3.8]. □

The following corollary shows that if a function has a positive derivative, then that function preserves the direction—in other words, the orientation—on the real line.

Corollary 1. *Under the hypotheses of* **Theorem 1**, *if $f'(x) > 0$ at every point of the open interval $]a, b[$, then f increases: if $a \leq v < w \leq b$, then $f(v) < f(w)$.*

Proof: Apply the theorem to the interval $[v, w]$. There is a point $x \in]v, w[$ such that $f(w) - f(v) = (w - v)\, f'(x) > 0$ because $(w - v) > 0$ and $f'(x) > 0$. □

The exercises that follow demonstrate other qualitative uses of the derivative, which anticipate similar uses with several variables, such as maps that preserve angles, areas, or orientation.

Exercises

The following four exercises demonstrate how tangent lines may outline the shape of a function.

1. Consider the function q defined by $q(x) = x^2$.

a) On graph paper or on a computer screen, plot the seven points $(-3/2, q(-3/2))$, $(-1, q(-1))$, $(-1/2, q(-1/2))$, $(0, q(0))$, $(1/2, q(1/2))$, $(1, q(1))$, and $(3/2, q(3/2))$.

b) Find a formula for $q'(x)$.

c) For each of the seven points already ploted on q, calculate the slope of the line tangent to q at that point, and graph a piece of that tangent line through that point on the same plot.

d) Draw q, or have a computer draw it.

2. Consider the function r defined by $r(x) = 1/x$.

a) On graph paper or on a computer screen, plot these four points: $(1/2, r(1/2))$, $(1, r(1))$, $(3/2, r(3/2))$, and $(2, r(2))$.

b) Find a formula for $r'(x)$.

c) For each of the four points already plotted on r, calculate the slope of the line tangent to r at that point, and graph a piece of that tangent line through that point on the same plot.

d) Draw r, or have a computer draw it.

3. On a common graph, plot both the function q defined by $q(x) = x^2$ and its tangent line $T_{q,1}$ at the point $(1, q(1))$, in the region where $0.9 \leq x \leq 1.1$.

4. On a common graph, plot both the function r defined by $r(x) = 1/x$ and its tangent line $T_{r,1}$ at the point $(1, r(1))$, in the region where $0.9 \leq x \leq 1.1$.

The following two exercises justify the definition of the derivative by **(1)**.

5. This exercise shows how the definition of the derivative distinguishes the tangent line from all other lines. Prove that

$$\lim_{h \to 0} \frac{f(x+h) - T_{f,x}(x+h)}{h} = 0, \quad \lim_{h \to 0} \frac{f(x+h) - L(x+h)}{h} \neq 0,$$

for every differentiable function f and every line L different from the tangent line $T_{f,x}$.

6. This exercise shows that without the quotient by h, the definition of the derivative would *not* distinguish the tangent line from all other lines: Prove that

$$\lim_{h \to 0} [f(x+h) - L(x+h)] = 0$$

for every continuous function f and every line L passing through $(x, f(x))$.

The following two exercises confirm that the differential remains invariant under translations (shifts) and homotheties (dilations).

7. Verify that a dilation of the plane by a constant multiplicative factor r does not affect the differential of a function. That is, for each real constant $r \neq 0$, define $g(x) = r\, f(r^{-1}x)$ for every $x \in \mathcal{D}$; then verify that $Dg_x = Df_{r^{-1}x}$.

8. Verify that a translation by a constant does not affect the differential of a function. That is, for each constant $c \in \mathbb{R}$, define $g(x) = f(x) + c$ for every $x \in \mathcal{D}$; then verify that $Dg_x = Df_x$.

The following two exercises identify the functions that preserve distances.

9. Assume that f is defined, but not necessarily supposed continuous, at every point of the entire real line $\mathbb{R}$, and assume also that f preserves distances between points: $|f(w) - f(v)| = |w - v|$ for all points $v, w \in \mathbb{R}$. Prove that three is a constant $c \in \mathbb{R}$ such that either $f(x) = c + x$ for every $x \in \mathbb{R}$ or $f(x) = c - x$ for every $x \in \mathbb{R}$. Thus, the only functions that preserve all distances on the real line are the identity $x \mapsto x$ and the central symmetry about the origin $x \mapsto -x$, followed by a shift (translation) by c.

10. Prove that if f is continuous at every point of the closed interval $[a, b]$, and if at every point of the open interval $]a, b[$ either $f'(x) = 1$ or $f'(x) = -1$, then f preserves distances between points; that is, for all points $v, w \in [a, b]$, $|f(w) - f(v)| = |w - v|$. Conclude that f' remains constant: Either $f'(x) = 1$ at every $x \in]a, b[$ or $f'(x) = -1$ at every $x \in]a, b[$.

2.2 The Differential of Planar Curves

Recall that a *curve* in the plane is a continuous function $\vec{\mathbf{f}} : \mathcal{D} \subseteq \mathbb{R} \to \mathbb{R}^2$ from an interval $\mathcal{D}$ on the real line $\mathbb{R}$ to the plane $\mathbb{R}^2$. Thus, for each $t \in \mathcal{D}$, the value $\vec{\mathbf{f}}(t)$ has two coordinates, denoted here by $f_1(t)$ and $f_2(t)$:

$$\vec{\mathbf{f}}(t) = \begin{pmatrix} f_1(t) \\ f_2(t) \end{pmatrix}.$$

The *derivative* $\vec{\mathbf{f}}'$ of such a curve $\vec{\mathbf{f}}$ admits a definition similar to the derivative of a function $f : \mathbb{R} \to \mathbb{R}$:

$$\begin{aligned} \vec{\mathbf{f}}'(t) &= \lim_{h \to 0} \frac{\vec{\mathbf{f}}(t+h) - \vec{\mathbf{f}}(t)}{h} \\ &= \lim_{h \to 0} \frac{1}{h} \left[\begin{pmatrix} f_1(t+h) \\ f_2(t+h) \end{pmatrix} - \begin{pmatrix} f_1(t) \\ f_2(t) \end{pmatrix} \right] \\ &= \lim_{h \to 0} \begin{pmatrix} [f_1(t+h) - f_1(t)]/h \\ [f_2(t+h) - f_2(t)]/h \end{pmatrix}, \end{aligned}$$

provided that the limits exist. From the definition of limits in the plane, it follows that

$$\vec{\mathbf{f}}'(t) = \begin{pmatrix} f_1'(t) \\ f_2'(t) \end{pmatrix}.$$

Thus, the coordinates of the derivative of a curve are the derivatives of the coordinates of that curve: $\vec{\mathbf{f}}' = (f_1', f_2')$. The vector $\vec{\mathbf{f}}'(t)$ is also called the *tangent vector* at t.

Application 2. If $\vec{\mathbf{f}}(t)$ represents the position at time t of an object in the plane, then $[\vec{\mathbf{f}}(t+h) - \vec{\mathbf{f}}(t)]/h$ represents the average speed, in direction and magnitude, of the object in the interval of time from t to $t+h$. The limit $\vec{\mathbf{f}}'(t)$ represents the speed of the object at time t. □

Remark 1. Because of the identification of the Euclidean plane with $\mathbb{R}^2$, no difference exists between points and vectors. Such an identification alleviates the need to consider vectors as equivalence classes of pairs of points. A similar remark applies to the identification of the Euclidean space with $\mathbb{R}^3$. □

Yet other interpretations of the derivative exist. For instance, with the absolute value $|\;|$ replaced by the Euclidean norm $\|\;\|$, defined by

$$\|(x_1, \ldots, x_n)\| \equiv \sqrt{x_1^2 + \cdots + x_n^2},$$

the limit that defines the derivative means that for each positive number ε, there is a positive number δ for which

$$\left\| \frac{\vec{\mathbf{f}}(t+h) - \vec{\mathbf{f}}(t)}{h} - \vec{\mathbf{f}}'(t) \right\| < \varepsilon$$

for every h such that $0 < |h| < \delta$, or, equivalently,

$$\left\| \vec{\mathbf{f}}(t+h) - \left\{ \vec{\mathbf{f}}(t) + h\,\vec{\mathbf{f}}'(t) \right\} \right\| < \varepsilon |h|,$$

for every h such that $|h| < \delta$. As for a function $f : \mathbb{R} \to \mathbb{R}$, the inequality just obtained means that the line $\vec{\mathbf{T}}_{\vec{\mathbf{f}},t} : \mathbb{R} \to \mathbb{R}^2$ with parametric equation

$$\vec{\mathbf{T}}_{\vec{\mathbf{f}},t}(t+h) = \vec{\mathbf{f}}(t) + h\,\vec{\mathbf{f}}'(t)$$

remains closer to the curve $\vec{\mathbf{f}}$ than does any other line $\vec{\mathbf{L}}$ in some neighborhood of the point $(t, \vec{\mathbf{f}}(t))$. Therefore, the line $\vec{\mathbf{T}}_{\vec{\mathbf{f}},t}$ is also called the line *tangent* to $\vec{\mathbf{f}}$ at t (or "at $(t, \vec{\mathbf{f}}(t))$").

Also, the *differential* of a curve $\vec{\mathbf{f}}$ at t is the function $D\vec{\mathbf{f}}_t$ defined by the formula

$$D\vec{\mathbf{f}}_t(h) = h\,\vec{\mathbf{f}}'(t).$$

Thus, the differential of a curve is the linear part of the parametric form of the tangent line with tangent vector $\vec{\mathbf{f}}'(t)$. As for functions into the real line, the differential of a planar curve remains invariant under translations and dilations.

Example 1. The function $\vec{\mathbf{L}}: [0,1] \to \mathbb{R}^2$ defined by

$$\vec{\mathbf{L}}(t) = (1-t)\begin{pmatrix} x_0 \\ y_0 \end{pmatrix} + t\begin{pmatrix} x_1 \\ y_1 \end{pmatrix} = \begin{pmatrix} (1-t)x_0 + tx_1 \\ (1-t)y_0 + ty_1 \end{pmatrix}$$

traces the straight line segment from the point $P_0 = (x_0, y_0)$ to the point $P_1 = (x_1, y_1)$, both included, with a constant tangent vector pointing from (x_0, y_0) to (x_1, y_1):

$$\vec{\mathbf{L}}'(t) = \begin{pmatrix} x_1 \\ y_1 \end{pmatrix} - \begin{pmatrix} x_0 \\ y_0 \end{pmatrix} = \begin{pmatrix} x_1 - x_0 \\ y_1 - y_0 \end{pmatrix}.$$

Thus, the differential of a line $\vec{\mathbf{L}}$ at t is the function $D\vec{\mathbf{L}}_t$ with the formula $D\vec{\mathbf{L}}_t(h) = h \cdot \vec{\mathbf{L}}'(t)$, which represents a straight line parallel to $\vec{\mathbf{L}}$ but passing through the origin. □

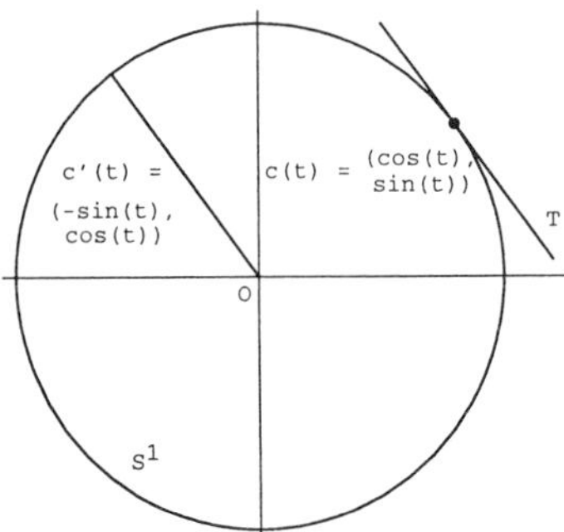

Figure 2. The unit circle with a piece of the line tangent at one point.

Example 2. The function $\vec{\mathbf{c}}: [0, 2\pi] \to \mathbb{R}^2$ defined by

$$\vec{\mathbf{c}}(t) = \begin{pmatrix} \cos(t) \\ \sin(t) \end{pmatrix}$$

traces the *unit circle* $S^1 \equiv \{(x,y) \in \mathbb{R}^2 : x^2 + y^2 = 1\}$ with symbolically concise but computationally demanding trigonometric functions, as shown in **Figure 2**. The derivative takes the form

$$\vec{\mathbf{c}}'(t) = \begin{pmatrix} \cos'(t) \\ \sin'(t) \end{pmatrix} = \begin{pmatrix} -\sin(t) \\ \cos(t) \end{pmatrix}.$$

As a verification, the usual dot product $\langle\, , \,\rangle$ shows that $\langle \vec{\mathbf{c}}(t), \vec{\mathbf{c}}'(t) \rangle = 0$, which confirms that the tangent line lies perpendicularly to the radius. Thus, the line tangent to the unit circle at the point $\vec{\mathbf{c}}(t) = (\cos(t), \sin(t))$ has the parametric form

$$\vec{\mathbf{T}}_{\vec{\mathbf{c}},t}(t+h) = \vec{\mathbf{c}}(t) + h \cdot \vec{\mathbf{c}}'(t) = \begin{pmatrix} \cos(t) \\ \sin(t) \end{pmatrix} + h \cdot \begin{pmatrix} -\sin(t) \\ \cos(t) \end{pmatrix}$$

with the differential of $\vec{\mathbf{c}}$ consisting only of the linear part:

$$D\vec{\mathbf{c}}_t(h) = h\,\vec{\mathbf{c}}'(t) = h\begin{pmatrix} -\sin(t) \\ \cos(t) \end{pmatrix}. \qquad \square$$

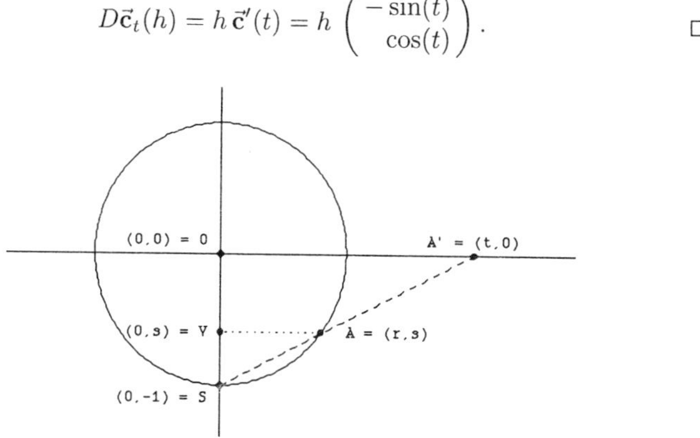

Figure 3. The inverse stereographic projection from the South Pole maps the point $t \in \mathbb{R}$ to the point $(r, s) \in S^1$.

Example 3. The inverse of the *stereographic projection* from the South Pole $S = (0, -1)$,

$$\vec{\mathbf{s}} :]-\infty, \infty[\to \mathbb{R}^2, \quad \vec{\mathbf{s}}(t) = \begin{pmatrix} 2t/(1+t^2) \\ (1-t^2)/(1+t^2) \end{pmatrix},$$

traces the unit circle without the South Pole S by means of computationally simple rational functions, established by means of similar triangles SOA' and SVA in **Figure 3**. Indeed,

$$r/(s - [-1]) = \overline{VA}/\overline{SV} = \overline{OA'}/\overline{SO} = t/1,$$

whence $r = (s+1)t$. Substituting $r = (s+1)t$ into the equation of the unit circle $r^2 + s^2 = 1$ produces $(s+1)^2 t^2 + s^2 = 1$, which simplifies to $(t^2+1)s^2 + 2t^2 s + (t^2 - 1) = 0$. Consequently, the quadratic formula gives $s = (-t^2 \pm \sqrt{t^4 - (t^2+1)(t^2-1)})/(t^2+1) = (-t^2 \pm 1)/(t^2+1)$. The solution $s = (-t^2-1)/(t^2+1) = -1$ corresponds to the South Pole, while the solution $s = (-t^2+1)/(t^2+1)$ yields the second coordinate of the other intersection at $A = (r, s)$ in **Figure 3**. Finally, returning to $r = (s+1)t$ gives $r = (s+1)t = t\{[(-t^2+1)/(t^2+1)] + [(t^2+1)/(t^2+1)]\} = 2t/(t^2+1)$.

As a verification that $\vec{s}(t)$ lies on the unit circle, the dot product gives $\|\vec{s}(t)\|^2 = \langle \vec{s}(t), \vec{s}(t) \rangle = 1$. For the derivative, the quotient rule yields the expression

$$\begin{aligned} \vec{\mathbf{s}}'(t) &= \begin{pmatrix} 2t/(1+t^2) \\ (1-t^2)/(1+t^2) \end{pmatrix}' \\ &= \begin{pmatrix} [2(1+t^2) - 2t(2t)]/(1+t^2)^2 \\ [-2t(1+t^2) - (1-t^2)(2t)]/(1+t^2)^2 \end{pmatrix} \end{aligned}$$

$$= \begin{pmatrix} 2(1-t^2)/(1+t^2)^2 \\ -4t/(1+t^2)^2 \end{pmatrix}.$$

As another verification, the dot product shows that $\langle \vec{s}(t), \vec{s}\,'(t) \rangle = 0$, which confirms that the tangent line lies perpendicularly to the radius.

□

Application 3. Steregraphic projections have many applications. For instance, they can be used to find the intersections of any two conic sections in the plane, such as the intersections of two hyperbolae in the Long Range Navigation (LORAN) system. The method consists first in parametrizing either of the two conic sections with an inverse stereographic projection, as in **Example 3**, which involves only ratios of quadratic polynomials [Reid 1990, 10–16; Shafarevich 1977, 6]. The method then proceeds by substituting the resulting quadratic rational formulae into the quadratic equation of the other conic section, which produces one quartic equation with t as the only variable.

Exercises

11. This exercise justifies the calculation of derivatives of curves through the derivatives of the coordinates. From the definition of limits in the plane, verify that

$$\lim_{h\to 0} \begin{pmatrix} [f_1(t+h) - f_1(t)]/h \\ [f_2(t+h) - f_2(t)]/h \end{pmatrix} = \begin{pmatrix} \lim_{h\to 0}[f_1(t+h) - f_1(t)]/h \\ \lim_{h\to 0}[f_2(t+h) - f_2(t)]/h \end{pmatrix} = \begin{pmatrix} f_1'(t) \\ f_2'(t) \end{pmatrix}.$$

The following six exercises investigate various qualitative features of the derivatives.

12. Verify that a translation by a constant vector does not affect the differential of a function. That is, for each vector $\vec{k} \in \mathbb{R}^2$, define $\vec{g}(t) = \vec{f}(t) + \vec{k}$, and prove that $D\vec{g}_t = D\vec{f}_t$.

13. Verify that the line $\vec{T}_{\vec{f},t}$ tangent to a curve $\vec{f}$ at t remains closer to $\vec{f}$ than does any other line $\vec{L}$ in a neighborhood of $(t, \vec{f}(t))$.

14. Consider again the curve $\vec{c}$ that traces the unit circle S^1 according to the formula $\vec{c}(t) = (\cos(t), \sin(t))$. For each $s \neq 0$, verify that the curve $\vec{f}$ defined by the formula $\vec{f}(t) = \vec{c}(s \cdot t) = (\cos(s \cdot t), \sin(s \cdot t))$ also traces the unit circle, but with a speed of magnitude $|s|$. Then interpret the significance of the sign of s: How do curves with $s > 0$ differ from curves with $s < 0$?

15. Verify that the inverse stereographic projection $\vec{s}$ does *not* preserve distances between the real line and the unit circle. That is, find two real numbers v and w for which $|w - v|$ does *not* equal the length of the arc from $\vec{s}(v)$ to $\vec{s}(w)$.

16. Consider again the curve $\vec{c}$ that traces the unit circle S^1 according to the formula $\vec{c}(t) = (\cos(t), \sin(t))$. Verify that for all points such that $u < v$ in $[0, 2\pi]$, the arc of the unit circle from $\vec{c}(v)$ to $\vec{c}(w)$ has the same length as the segment $[v, w]$. Thus, $\vec{c}$ preserves distances between the real line and the unit circle, with distances measured not through the plane but along the unit circle.

17. Prove that if a curve $\vec{f} : \mathbb{R} \to \mathbb{R}^2$ has a continuous derivative $\vec{f}'$ and if $\vec{f}$ preserves distances—in the sense that for every pair of real numbers v and w, the distance $|v - w|$ equals the distance measured along the arc of the curve $\vec{f}$ from $\vec{f}(v)$ to $\vec{f}(w)$—then $\|\vec{f}'(t)\| = 1$ for every $t \in \mathbb{R}$.

The following three exercises provide some preparation for stereographic map projections.

18. Consider the function $\vec{n} : \mathbb{R} \to S^1$ defined by the following geometric construction. That is, for each real number $t \in \mathbb{R}$, the point $\vec{n}(t) = (n_1(t), n_2(t))$ lies at the second intersection of the unit circle S^1 and the straight line L through $(0, 1)$ and $(t, 0)$. (The first intersection is the North Pole $(0, 1)$.) Find algebraic formulae for $n_1(t)$ and $n_2(t)$.

19. With the inverse stereographic projection from the South Pole, $\vec{s}$, described in **Example 3**, establish a formula for the number t corresponding to the point $(x, z) \in S^1$ by solving for t the equations $\vec{s}(t) = (x, z)$, subject to the restriction that $x^2 + z^2 = 1$.

20. With the inverse stereographic projection from the North Pole, $\vec{n}$, described in **Exercise 18**, establish a formula for the number t corresponding to the point $(x, z) \in S^1$ by solving for t the equations $\vec{n}(t) = (x, z)$, subject to the restriction that $x^2 + z^2 = 1$.

The following two exercises point to other geometric applications of planar curves.

21. Consider a parabola C with focus F and directrix L. For each point P on the parabola C, prove that the line T tangent to C at P makes equal angles with the directrix and the line from F to P.

Application 4. A parabolic mirror reflects all light rays emerging from the focus into the direction of the directrix, thus emitting a parallel beam. Conversely, a parabolic mirror concentrates at the focus all light rays coming parallelly to the directrix. □

22. Some curves have unexpected applications. Verify that for each *rational* number $t = p/q \in \mathbb{Q}$, the inverse stereographic projection $\vec{s}$ defined in **Example 3** gives a point with *rational* coordinates on the unit circle; moreover, clearing denominators, verify that such a rational point then yields a Pythagorean triple (m, n, r) of three *integers* such that $m^2 + n^2 = r^2$. Finally, verify that the construction just described yields *all* the Pythagorean triples.

Application 5. Pythagorean triples have served for thousands of years to construct pedagogical examples of mathematical problems so that square roots do not lead to irrational numbers [van der Waerden 1983, 33], and now they also allow for test cases to test the accuracy of some computer programs independently from rounding errors. Pythagorean triples also occur in megalithic monuments and altars [van der Waerden 1983, 17–25]. □

2.3 The Differential of Spatial Curves

Similar to a curve in the plane, a curve in space is a continuous function $\vec{\mathbf{f}} : \mathcal{D} \subseteq \mathbb{R} \to \mathbb{R}^3$ from an interval $\mathcal{D}$ on the real line $\mathbb{R}$ to the three-dimensional space $\mathbb{R}^3$. Thus, for each $t \in \mathcal{D}$, the value $\vec{\mathbf{f}}(t)$ has three coordinates, denoted here by $f_1(t)$, $f_2(t)$, and $f_3(t)$:

$$\vec{\mathbf{f}}(t) = \begin{pmatrix} f_1(t) \\ f_2(t) \\ f_3(t) \end{pmatrix},$$

with derivative

$$\begin{aligned} \vec{\mathbf{f}}'(t) &= \lim_{h \to 0} \frac{\vec{\mathbf{f}}(t+h) - \vec{\mathbf{f}}(t)}{h} \\ &= \lim_{h \to 0} \begin{pmatrix} [f_1(t+h) - f_1(t)]/h \\ [f_2(t+h) - f_2(t)]/h \\ [f_3(t+h) - f_3(t)]/h \end{pmatrix} = \begin{pmatrix} f_1'(t) \\ f_2'(t) \\ f_3'(t) \end{pmatrix}, \end{aligned}$$

provided that the limits exist.

Application 6. If $\vec{\mathbf{f}}(t)$ represents the position at time t of an object in space, then $[\vec{\mathbf{f}}(t+h) - \vec{\mathbf{f}}(t)]/h$ represents the average speed, in direction and magnitude, of the object in the interval of time from t to $t+h$, and the limit $\vec{\mathbf{f}}'(t)$ represents the speed of the object at time t. □

As for planar curves, the line *tangent* to the spatial curve $\vec{\mathbf{f}}$ at the point $(t, \vec{\mathbf{f}}(t))$ is the line $\vec{\mathbf{T}}_{\vec{\mathbf{f}},t}$ parametrized by

$$\vec{\mathbf{T}}_{\vec{\mathbf{f}},t}(t+h) = \vec{\mathbf{f}}(t) + h\,\vec{\mathbf{f}}'(t).$$

In a neighborhood of the point of tangency, $(t, \vec{\mathbf{f}}(t))$, the tangent line $\vec{\mathbf{T}}_{\vec{\mathbf{f}},t}$ remains closer to the curve $\vec{\mathbf{f}}$ than does any other line. Also, the *differential* of a spatial curve $\vec{\mathbf{f}}$ at t is the function $D\vec{\mathbf{f}}_t$ defined by the formula

$$D\vec{\mathbf{f}}_t(h) = h\,\vec{\mathbf{f}}'(t).$$

Thus, the differential of a curve is the linear part of the parametric form of the tangent line with tangent vector $\vec{\mathbf{f}}'$. As for functions into the real line and for planar curves, the differential of a spatial curve remains invariant under translations and dilations.

Example 4. The function $\vec{\mathbf{L}}: [0,1] \to \mathbb{R}^3$ defined by

$$\vec{\mathbf{L}}(t) = (1-t)\begin{pmatrix} x_0 \\ y_0 \\ z_0 \end{pmatrix} + t\begin{pmatrix} x_1 \\ y_1 \\ z_1 \end{pmatrix} = \begin{pmatrix} (1-t)x_0 + tx_1 \\ (1-t)y_0 + ty_1 \\ (1-t)z_0 + tz_1 \end{pmatrix}$$

traces the straight line segment from the point $P_0 = (x_0, y_0, z_0)$ to the point $P_1 = (x_1, y_1, z_1)$, both included, with a constant tangent vector pointing from (x_0, y_0, z_0) to (x_1, y_1, z_1):

$$\vec{\mathbf{L}}'(t) = \begin{pmatrix} x_1 \\ y_1 \\ z_1 \end{pmatrix} - \begin{pmatrix} x_0 \\ y_0 \\ z_0 \end{pmatrix} = \begin{pmatrix} x_1 - x_0 \\ y_1 - y_0 \\ z_1 - z_0 \end{pmatrix}.$$

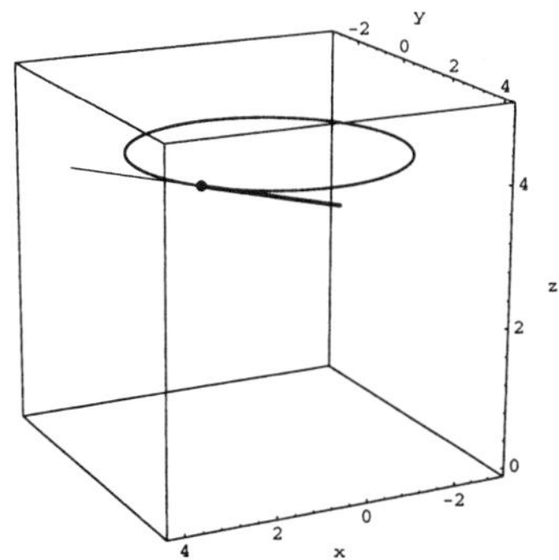

Figure 4. The parallel with radius $r = 3$ at height $H = 4$, displayed here with a piece of the line tangent at the point $(12/5, 9/5, 4)$.

Example 5. For each positive number r, the function $\vec{\mathbf{p}}: [0, 2\pi] \to \mathbb{R}^3$ defined by

$$\vec{\mathbf{p}}(t) = \begin{pmatrix} r\cos(t) \\ r\sin(t) \\ H \end{pmatrix}$$

traces the horizontal circle with radius r at height H, in the plane where $z = H$: $\{(x,y,z) \in \mathbb{R}^3 : x^2 + y^2 = r^2 \text{ and } z = H\}$, as shown in **Figure 4.** Its derivative takes the form

$$\vec{\mathbf{p}}'(t) = \begin{pmatrix} r\cos'(t) \\ r\sin'(t) \\ H' \end{pmatrix} = \begin{pmatrix} -r\sin(t) \\ r\cos(t) \\ 0 \end{pmatrix}.$$

For any $R > r$ and with the particular choice $H = \sqrt{R^2 - r^2}$, so that $H^2 + r^2 = R^2$, such a horizontal circle lies on the sphere with radius R

centered at the origin:

$$\begin{aligned}
\|\vec{\mathbf{p}}(t)\|^2 &= p_1^2(t) + p_2^2(t) + p_3^2(t) \\
&= (r\,\cos(t))^2 + (r\,\sin(t))^2 + H^2 \\
&= r^2\left([\cos(t)]^2 + [\sin(t)]^2\right) + H^2 = r^2 + H^2 = R^2.
\end{aligned}$$

Such circles form the *parallels* on the surface of the Earth. The circle with radius $r = \sqrt{R^2 - H^2}$ at height H on the sphere with radius R then represents the circle at latitude $\lambda = \text{Arcsin}\,(H/R)$. The *latitude* of a point $\vec{\mathbf{x}} \in \mathbb{R}^3$ is the oriented angle $\lambda \in [-\pi/2, \pi/2]$ from the projection $(x, y, 0)$ of $\vec{\mathbf{x}} = (x, y, z)$ on the horizontal coordinate plane $\mathbb{R}^2 \times \{0\}$ to $\vec{\mathbf{x}}$; thus, $\lambda = \text{Arctan}\,(z/\sqrt{x^2 + y^2})$. □

Example 6. For each positive number R, the function $\vec{\mathbf{q}}:\ [0, 2\pi] \to \mathbb{R}^3$ defined by

$$\vec{\mathbf{q}}(t) = \begin{pmatrix} R\,\cos(t) \\ 0 \\ R\,\sin(t) \end{pmatrix}$$

traces the vertical unit circle in the plane where $y = 0$: $\{(x, y, z) \in \mathbb{R}^3 : x^2 + z^2 = R^2 \text{ and } y = 0\}$, with derivative

$$\vec{\mathbf{q}}'(t) = \begin{pmatrix} R\,\cos'(t) \\ 0' \\ R\,\sin'(t) \end{pmatrix} = \begin{pmatrix} -R\,\sin(t) \\ 0 \\ R\,\cos(t) \end{pmatrix}.$$ □

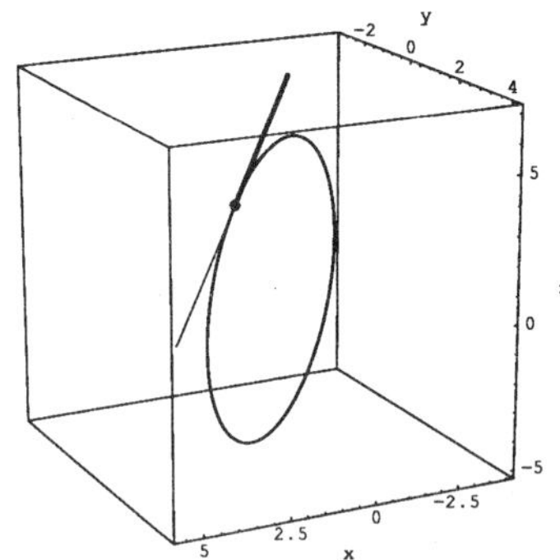

Figure 5. The meridian with radius $R = 5$ at longitude $\varphi = \text{Arccos}(4/5)$, displayed here with a piece of the line tangent at the point $(12/5, 9/5, 4)$.

Example 7. Rotated around the vertical axis, the circles in **Example 6** form the *meridians* on the surface of the Earth. To obtain such a meridian, multiply the vector $\vec{\mathbf{q}}(t)$ by the matrix Q of a rotation by an angle φ about the third (vertical) coordinate axis, as shown in **Figure 5**:

$$Q = \begin{pmatrix} \cos(\varphi) & -\sin(\varphi) & 0 \\ \sin(\varphi) & \cos(\varphi) & 0 \\ 0 & 0 & 1 \end{pmatrix},$$

$$\begin{aligned} Q\vec{\mathbf{q}}(t) &= \begin{pmatrix} \cos(\varphi) & -\sin(\varphi) & 0 \\ \sin(\varphi) & \cos(\varphi) & 0 \\ 0 & 0 & 1 \end{pmatrix} \begin{pmatrix} R\cos(t) \\ 0 \\ R\sin(t) \end{pmatrix} \\ &= \begin{pmatrix} \cos(\varphi)R\cos(t) - \sin(\varphi)\cdot 0 + 0\cdot R\sin(t) \\ \sin(\varphi)R\cos(t) + \cos(\varphi)\cdot 0 + 0\cdot R\sin(t) \\ 0\cdot R\cos(t) + 0\cdot 0 + 1\cdot R\sin(t) \end{pmatrix} \\ &= \begin{pmatrix} R\cos(\varphi)\cos(t) \\ R\sin(\varphi)\cos(t) \\ R\sin(t) \end{pmatrix}, \end{aligned}$$

which represents the meridian at *longitude* φ, with tangent vector

$$(Q\vec{\mathbf{q}})'(t) = \begin{pmatrix} -R\cos(\varphi)\sin(t) \\ -R\sin(\varphi)\sin(t) \\ R\cos(t) \end{pmatrix}.$$

For future reference, observe that the tangent vector points toward the North Pole. □

Remark 2. By a convention from the first International Meridian Conference on 22 October 1884, *longitudes* agree with mathematical angles in the equatorial plane viewed from the North, measured counterclockwise—in other words, with positive longitudes for the East and negative longitudes for the West, from the center of the transit instrument at the Observatory of Greenwich, in England [Brown 1979, 297; Snyder 1987b, 9].

Remark 3. Mathematically, two different functions $\vec{\mathbf{f}} : \mathcal{I} \subseteq \mathbb{R} \to \mathbb{R}$ and $\vec{\mathbf{g}} : \mathcal{J} \subseteq \mathbb{R} \to \mathbb{R}$ are also different curves, even if their images coincide. In other words, images of curves such as **Figures 2**, **4**, and **5** display only $\vec{\mathbf{f}}(\mathcal{I})$ or $\vec{\mathbf{g}}(\mathcal{J})$ without any indication of which number $t \in \mathcal{I}$ corresponds to which point $\vec{\mathbf{f}}(t)$ on the image. Thus, two different functions may give the same picture. In yet other words, the image only shows a track, whereas the function $\vec{\mathbf{f}}$ also indicates the "time schedule" at which it reaches each point on the track. □

Example 8. The inverse of the *stereographic projection* from the point $P = (-1, 0, 0)$ on the equator,

$$\vec{\mathbf{s}} :]-\infty, \infty[\to \mathbb{R}^3, \quad \vec{\mathbf{s}}(t) = \begin{pmatrix} 2t/(1+t^2) \\ 0 \\ (1-t^2)/(1+t^2) \end{pmatrix},$$

traces the vertical unit circle without the point P in the plane where $y = 0$: $\{(x, y, z) \in \mathbb{R}^3 : x^2 + z^2 = 1 \text{ and } y = 0\}$, with derivative

$$\begin{aligned} \vec{\mathbf{s}}'(t) &= \begin{pmatrix} [2(1+t^2) - 2t(2t)]/(1+t^2)^2 \\ 0 \\ [-2t(1+t^2) - (1-t^2)(2t)]/(1+t^2)^2 \end{pmatrix} \\ &= \begin{pmatrix} 2(1-t^2)/(1+t^2)^2 \\ 0 \\ -4t/(1+t^2)^2 \end{pmatrix}. \end{aligned}$$

Composed with rotations, such a stereographic projection gives a parametrization of the same meridians as those in **Examples 6** and **7**, but with one point removed and with a different parametrization—in other words, with a different schedule. □

Exercises

23. Prove that on each sphere, every meridian intersects every parallel.

24. By differentiating $\|\vec{\mathbf{f}}\|^2$, prove that if the image of a curve $\vec{\mathbf{f}}$ lies on the sphere $\mathcal{S}^2(0, R)$—in other words, if $\vec{\mathbf{f}}(t) \in \mathcal{S}^2(0, R)$ for every t in some open interval—then the vector $\vec{\mathbf{f}}'(t)$ tangent to the curve points in a direction perpendicular to the position vector $\vec{\mathbf{f}}(t)$ for every t in that interval: $\vec{\mathbf{f}}(t) \perp \vec{\mathbf{f}}'(t)$. (The symbol $\perp$ means "is perpendicular to.")

25. For each spatial curve $\vec{\mathbf{f}}$ and each constant $b \in \mathbb{R} \setminus \{0, 1\}$, consider the parametric line $\vec{\mathbf{L}} : \mathbb{R} \to \mathbb{R}^3$ defined by $\vec{\mathbf{L}}(t + h) = \vec{\mathbf{f}}(t) + b\,h\,\vec{\mathbf{f}}'(t)$. Verify that $\vec{\mathbf{L}}$ also parametrizes the line $\vec{\mathbf{T}}_{\vec{\mathbf{f}},t}$ tangent to $\vec{\mathbf{f}}$ at t, but with a different "speed," so that if $\vec{\mathbf{f}}'(t) \neq \vec{\mathbf{0}}$, then $\left\|\vec{\mathbf{f}}(t+h) - \vec{\mathbf{L}}(t+h)\right\| > \left\|\vec{\mathbf{f}}(t+h) - \vec{\mathbf{T}}_{\vec{\mathbf{f}},t}(t+h)\right\|$ for every h such that $0 < |h| < \delta$, for some appropriate positive δ.

26. The *Arctic Circle* is the parallel at latitude $66°33'$ (North), and the *Antarctic Circle* is the parallel at latitude $-66°33'$ (South). Both circles pertain to the inclination, by $23°27'$, of the Earth's axis relative to the plane of the Earth's orbit, called the *ecliptic*.

a) On the Summer Solstice (about June 21 or 22 in the Northern Hemisphere), light rays arriving from the sun intersect the South-North axis of the Earth at an angle of $90° + 66°33' = 156°33'$. Prove that on the Summer Solstice, all points at latitudes greater than $66°33'$ remain in the sunlight, regardless of their longitudes, a phenomenon called the *midnight sun*, while all points at latitudes smaller than $-66°33'$ remain in the dark, regardless of their longitudes.

b) On the Winter Solstice (about December 21 or 22 in the Northern Hemisphere), light rays arriving from the sun intersect the South-North axis of the Earth at an angle of $66°33'$. Prove that on the Winter Solstice, all points at latitudes greater than $66°33'$ remain in the dark, regardless of their longitudes, while all points at latitudes smaller than $-66°33'$ remain in the sun light, regardless of their longitudes.

2.4 Great Circles

Consider the sphere $\mathcal{S}^2(0, R)$ with radius R centered at the origin in $\mathbb{R}^3$:

$$\mathcal{S}^2(0, R) = \{(x, y, z) \in \mathbb{R}^3 : x^2 + y^2 + z^2 = R^2\}.$$

(The superscript ^{2} indicates that the sphere is a two-dimensional surface, with only two independent directions on it, for instance, North-South and East-West, even though the sphere lies in a three-dimensional space.) Such a sphere may represent the Earth or any other planet. Between any two points $\vec{\mathbf{A}}$ and $\vec{\mathbf{B}}$ on such a sphere, the shortest path that follows the surface consists of a *great circle*, which has the same radius as the sphere and lies in the plane P containing the two points $\vec{\mathbf{A}}$ and $\vec{\mathbf{B}}$ and the origin $\vec{\mathbf{0}}$, as demonstrated informally in reference [Vest and Benge 1994–1995]. Because the plane P passes through $\vec{\mathbf{0}}$, $\vec{\mathbf{A}}$, and $\vec{\mathbf{B}}$, the cross product $(n_1, n_2, n_3) = \vec{\mathbf{N}} = \vec{\mathbf{B}} \times \vec{\mathbf{A}}$ is perpendicular to the plane P of the great circle. As the intersection of the plane P and the sphere $\mathcal{S}^2(0, R)$, the great circle must satisfy the following two simultaneous equations:

$$\begin{cases} x^2 + y^2 + z^2 = R^2 & \text{equation of the sphere,} \\ n_1 x + n_2 y + n_3 z = 0 & \text{equation of the plane.} \end{cases}$$

Such a representation gives a great circle as the intersection of two surfaces, defined by two simultaneous equations. In some contexts, however, a parametric representation by a function $\vec{\mathbf{g}} : [0, 2\pi] \to \mathbb{R}^3$ may prove more convenient—for instance, to trace the great circle, by hand or with a computer. To obtain such a parametrization of a great circle in space, it suffices to adapt the unit circle in the plane, for instance, with an orthonormal basis in the plane P of the great circle instead of the usual orthonormal basis. Though the two distinct vectors $\vec{\mathbf{A}}$ and $\vec{\mathbf{B}}$ span the plane P, they need not be orthonormal; but the Gram-Schmidt algorithm or a cross product quickly produce the desired basis:

Algorithm 1. *A calculation of the great circle through two specified points on a sphere.*

Step 1. Compute the cross product $\vec{\mathbf{N}} = \vec{\mathbf{B}} \times \vec{\mathbf{A}}$.

Step 2. Compute the cross product $\vec{\mathbf{G}} = \vec{\mathbf{A}} \times \vec{\mathbf{N}}$.

Step 3. Compute the Euclidean norm $\|\vec{\mathbf{G}}\| = \sqrt{\langle \vec{\mathbf{G}}, \vec{\mathbf{G}} \rangle}$ and the unit vector $\vec{\mathbf{V}} = \vec{\mathbf{G}}/\|\vec{\mathbf{G}}\|$.

Step 4. Observing that $\|\vec{\mathbf{A}}\| = R$, compute the unit vector $\vec{\mathbf{U}} = \vec{\mathbf{A}}/\|\vec{\mathbf{A}}\| = \vec{\mathbf{A}}/R$.

The resulting unit vectors $\vec{\mathbf{U}}$ and $\vec{\mathbf{V}}$ form an orthonormal basis for the plane P with the same orientation as $\vec{\mathbf{A}}$ and $\vec{\mathbf{B}}$. Consequently, the following curve $\vec{\mathbf{g}} : [0, 2\pi] \to \mathbb{R}^3$ traces the great circle through $\vec{\mathbf{A}}$ and $\vec{\mathbf{B}}$ on the surface of the sphere, starting from $\vec{\mathbf{g}}(0) = \vec{\mathbf{A}}$ at $t = 0$:

$$\vec{\mathbf{g}}(t) = R\cos(t)\,\vec{\mathbf{U}} + R\sin(t)\,\vec{\mathbf{V}}.$$

To simplify computations, however, observe that $R\,\vec{\mathbf{U}} = \vec{\mathbf{A}}$. □

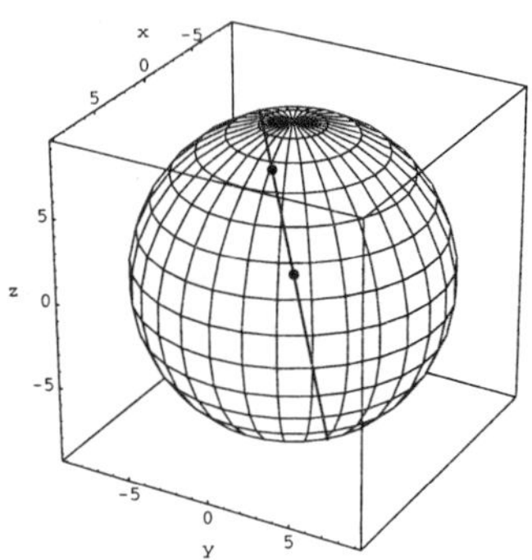

Figure 6. The great circle through $(4, 1, 8)$ and $(7, 4, 4)$ on the sphere $\mathcal{S}^2(0, 9)$.

Example 9. On the sphere $\mathcal{S}^2(0, 9)$ with radius $R = 9$ and center at the origin, consider the two points $\vec{\mathbf{A}} = (4, 1, 8)$ and $\vec{\mathbf{B}} = (7, 4, 4)$. Verify that they lie on the sphere by computing their Euclidean norms:

$$\|\vec{\mathbf{A}}\| = \|(4, 1, 8)\| = \sqrt{4^2 + 1^2 + 8^2} = \sqrt{16 + 1 + 64} = \sqrt{81} = 9,$$

$$\|\vec{\mathbf{B}}\| = \|(7, 4, 4)\| = \sqrt{7^2 + 4^2 + 4^2} = \sqrt{49 + 16 + 16} = \sqrt{81} = 9.$$

The algorithm gives the following orthonormal basis:

Step 1. $\vec{\mathbf{N}} = \vec{\mathbf{B}} \times \vec{\mathbf{A}} = (7, 4, 4) \times (4, 1, 8) = (28, -40, -9)$.
Step 2. $\vec{\mathbf{G}} = \vec{\mathbf{A}} \times \vec{\mathbf{N}} = (4, 1, 8) \times (28, -40, -9) = (311, 260, -188)$.
Step 3. $\|\vec{\mathbf{G}}\| = \sqrt{199{,}665}$ and $\vec{\mathbf{V}} = \vec{\mathbf{G}}/\|\vec{\mathbf{G}}\| = (311, 260, -188)/\sqrt{199{,}665}$.
Step 4. $\vec{\mathbf{U}} = \vec{\mathbf{A}}/\|\vec{\mathbf{A}}\| = \vec{\mathbf{A}}/R = (4, 1, 8)/9$.

Hence, the great circle through $(4,1,8)$ and $(7,4,4)$, shown in **Figure 6**, has the parametrization:

$$\begin{aligned}\vec{\mathbf{g}}(t) &= 9\cos(t)\,\frac{1}{9}\begin{pmatrix}4\\1\\8\end{pmatrix}+9\sin(t)\,\frac{1}{\sqrt{199{,}665}}\begin{pmatrix}311\\260\\-188\end{pmatrix}\\ &= \begin{pmatrix}4\cos(t)+9\times 311\sin(t)/\sqrt{199{,}665}\\ 1\cos(t)+9\times 260\sin(t)/\sqrt{199{,}665}\\ 8\cos(t)-9\times 188\sin(t)/\sqrt{199{,}665}\end{pmatrix}.\end{aligned}$$

□

Remark 4. A sphere provides only a first approximation of the shape of the Earth, which, like that of the planet Mars, more closely resembles an "oblate spheroid," or, in other words, an ellipsoid with two mutually equal equatorial principal axes and a slightly shorter polar principal axis. Moreover, some satellites of some of the planets more closely resemble ellipsoids with three different principal axes [Snyder 1987b, 11–12]. For such ellipsoids, however, the shortest path between two points on the surface may differ from a circle and may follow a much more complicated curve instead [Audin 1994]. Because the Earth's polar axis differs from the equatorial axis by only one part in 297, or about 0.34%, spheres and great circles suffice for the present purposes.

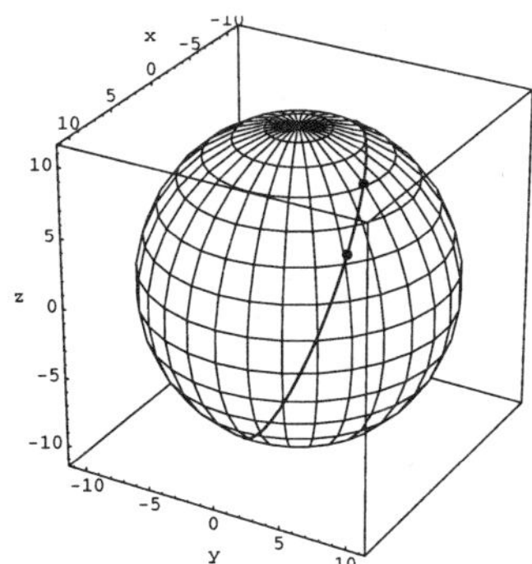

Figure 7. The great circle through $(2,6,9)$ and $(6,7,6)$ on the sphere $\mathcal{S}^2(0,11)$.

Exercises

27. Parametrize the great circle that passes through the points $(2,6,9)$ and $(6,7,6)$ on the sphere with radius 11 centered at the origin, as displayed in **Figure 7**.

28. Verify that if $t \mapsto \vec{\mathbf{g}}(t)$ parametrizes a great circle, then so do $t \mapsto \vec{\mathbf{g}}(-t)$ and $t \mapsto \vec{\mathbf{g}}(s \cdot t)$ for each nonzero $s \in \mathbb{R} \setminus \{0\}$.

29. Develop a method to parametrize the great circle through $\vec{\mathbf{A}}$ and $\vec{\mathbf{B}}$ with a specified speed v, so that $\|\vec{\mathbf{g}}'(s \cdot t)\| = v$. Such a method would then produce the position $\vec{\mathbf{g}}(s \cdot t)$ at time t of an object traveling at speed v from $\vec{\mathbf{A}}$ to $\vec{\mathbf{B}}$.

30. Because each great circle determines *two* arcs of that circle through $\vec{\mathbf{A}}$ and $\vec{\mathbf{B}}$ on it, usually only one of these two arcs is the shortest path from $\vec{\mathbf{A}}$ to $\vec{\mathbf{B}}$. Develop a method to modify the algorithm so that it guarantees that the curve $\vec{\mathbf{g}}$ starts from $\vec{\mathbf{A}}$ along the shortest arc to $\vec{\mathbf{B}}$.

31. At each instant, the *Terminator* is the curve T that separates day from night on the surface of the Earth [Edelman and Kostlan 1995, 4–5]. Under the assumptions of a spherical Earth and parallel light rays, prove that the Terminator is a great circle.

32. The *Tropic of Cancer* is the parallel at latitude $23°27'$ (North), and the *Tropic of Capricorn* is the parallel at latitude $-23°27'$ (South).

Recall from **Exercise 26** that the Arctic Circle is the parallel at latitude $66°33'$ (North), the Antarctic Circle is the parallel at latitude $-66°33'$ (South), and the inclination of the Earth's axis relatively to the plane of the Earth's orbit amounts to $23°27'$. Moreover, on the Summer Solstice, light rays arriving from the sun intersect the South-North axis of the Earth at an angle of $90° + 66°33' = 156°33'$, whereas on the Winter Solstice, light rays arriving from the sun intersect the South-North axis of the Earth at an angle of $66°33'$.

a) Prove that on either solstice, the plane of the Terminator is perpendicular to a vector from the center of the Earth to a point on a Tropic. (Which Tropic depends upon which solstice.)

b) Prove that on either solstice, the points with the highest and lowest latitudes on the Terminator lie at the intersections of the Terminator with the Arctic and Antarctic Circles.

2.5 Angles Between Curves in Euclidean Spaces

Freshman calculus texts show that two nonzero vectors $\vec{\mathbf{u}}$ and $\vec{\mathbf{v}}$, either in the plane or in space, form an angle θ with cosine expressed by the formula [Strang 1991, 404]

$$\cos(\theta) = \frac{\langle \vec{\mathbf{u}}, \vec{\mathbf{v}} \rangle}{\|\vec{\mathbf{u}}\| \cdot \|\vec{\mathbf{v}}\|}, \qquad \theta = \text{Arccos}\left(\frac{\langle \vec{\mathbf{u}}, \vec{\mathbf{v}} \rangle}{\|\vec{\mathbf{u}}\| \cdot \|\vec{\mathbf{v}}\|}\right),$$

where $\vec{\mathbf{u}} \cdot \vec{\mathbf{v}} = \sum_{i=1}^{n} u_i v_i$ and $\|\vec{\mathbf{u}}\| = \sqrt{\langle \vec{\mathbf{u}}, \vec{\mathbf{u}} \rangle}$, with $n = 2$ for the two-dimensional plane and $n = 3$ for the three-dimensional space.

Example 10. For the vectors $\vec{\mathbf{u}} = (2, 3, 6)$ and $\vec{\mathbf{v}} = (8, 5, 3)$,

$$\langle \vec{\mathbf{u}}, \vec{\mathbf{v}} \rangle = (2,3,6) \cdot (8,5,3) = (2 \cdot 8) + (3 \cdot 5) + (6 \cdot 3) = 49,$$

$$
\begin{aligned}
\|\vec{\mathbf{u}}\| &= \|(2,3,6)\| = \sqrt{2^2+3^2+6^2} = \sqrt{49} = 7, \\
\|\vec{\mathbf{v}}\| &= \|(8,5,3)\| = \sqrt{8^2+5^2+3^2} = \sqrt{98} = 7\sqrt{2}, \\
\cos(\theta) &= \frac{\langle \vec{\mathbf{u}}, \vec{\mathbf{v}} \rangle}{\|\vec{\mathbf{u}}\| \cdot \|\vec{\mathbf{v}}\|} = \frac{49}{7 \cdot 7\sqrt{2}} = \frac{1}{\sqrt{2}}, \\
\theta &= \operatorname{Arccos}\left(\frac{1}{\sqrt{2}}\right) = \frac{\pi}{4}.
\end{aligned}
$$
□

The angle between two vectors also equals the angle (interior or exterior angle, depending upon the orientations of the lines) between the two straight lines passing through the origin and containing the vectors. Hence, the concept of angle between two straight lines extends to that of angle between two curves, defined as the angle between the straight lines tangent to the curves that meet at a common point.

Definition 1. A *smooth curve* is a function $\vec{\mathbf{p}} : \,]a,b[\subset \mathbf{R} \to \mathbf{R}^n$, differentiable on an open interval $]a,b[$, whose derivative vanishes nowhere: $\vec{\mathbf{p}}'(t) \neq \vec{\mathbf{0}}$ for every $t \in]a,b[$. The *angle* between two smooth curves $\vec{\mathbf{p}}$ and $\vec{\mathbf{q}}$ that meet at a common point $\vec{\mathbf{p}}(t_p) = \vec{\mathbf{x}} = \vec{\mathbf{q}}(t_q)$ is the angle between their tangent vectors, $\vec{\mathbf{p}}'(t_p)$ and $\vec{\mathbf{q}}'(t_q)$.

Thus, two smooth curves $\vec{\mathbf{p}}$ and $\vec{\mathbf{q}}$ that meet at $\vec{\mathbf{p}}(t_p) = \vec{\mathbf{x}} = \vec{\mathbf{q}}(t_q)$ make at that point an angle θ given by the formula

$$\theta = \operatorname{Arccos}\left(\frac{\langle \vec{\mathbf{p}}'(t_p), \vec{\mathbf{q}}'(t_q) \rangle}{\|\vec{\mathbf{p}}'(t_p)\| \cdot \|\vec{\mathbf{q}}'(t_q)\|}\right).$$
□

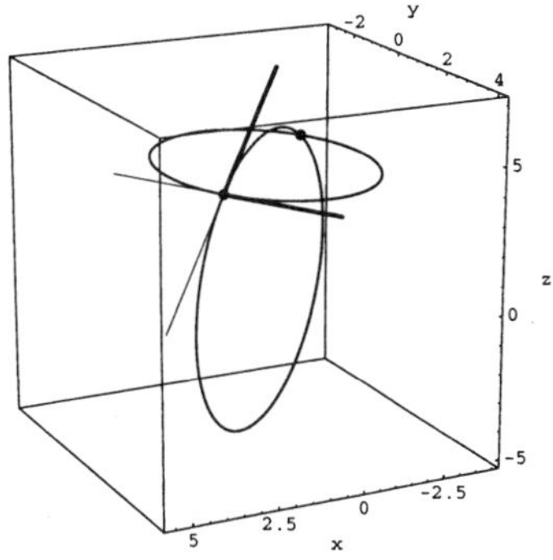

Figure 8. Meridians intersect parallels at a right angle.

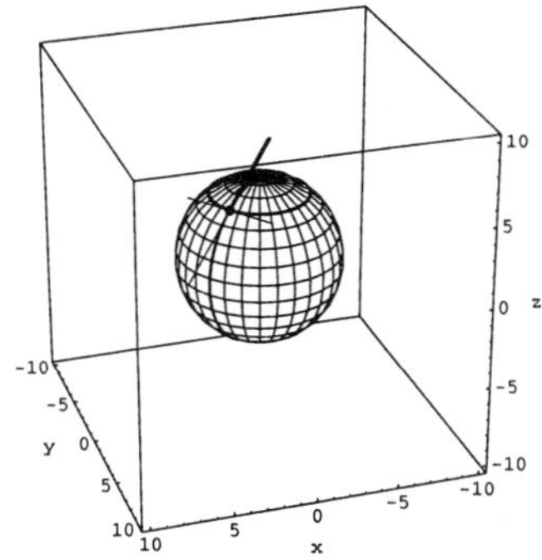

Figure 9. The same intersection with the underlying sphere.

Example 11. On the sphere $\mathcal{S}^2(0,R)$ with radius R, every parallel crosses every meridian at a right angle. That is, the parallel $\vec{\mathbf{p}} : [0, 2\pi] \to$

$\mathbb{R}^3$ at height H and latitude $\lambda = \text{Arcsin}\,(H/R)$ with radius $r = \sqrt{R^2 - H^2}$ has the formula

$$\vec{\mathbf{p}}(t) = \begin{pmatrix} r\,\cos(t) \\ r\,\sin(t) \\ H \end{pmatrix}, \quad \vec{\mathbf{p}}'(t) = \begin{pmatrix} -r\,\sin(t) \\ r\,\cos(t) \\ 0 \end{pmatrix},$$

whereas the meridian $\vec{\mathbf{q}} : [0, 2\pi] \to \mathbb{R}^3$ at longitude $-\varphi$ has the formula

$$\vec{\mathbf{q}}(t) = \begin{pmatrix} R\,\cos(\varphi)\cos(t) \\ R\,\sin(\varphi)\cos(t) \\ R\,\sin(t) \end{pmatrix}, \quad \vec{\mathbf{q}}'(t) = \begin{pmatrix} -R\,\cos(\varphi)\sin(t) \\ -R\,\sin(\varphi)\sin(t) \\ R\,\cos(t) \end{pmatrix}.$$

The parallel $\vec{\mathbf{p}}$ and the meridian $\vec{\mathbf{q}}$ intersect each other at the point

$$\vec{\mathbf{x}} = (R\,\cos(\varphi)\cos(\lambda), R\,\sin(\varphi)\cos(\lambda), R\,\sin(\lambda))$$

at longitude $-\varphi$ and latitude $\lambda = \text{Arcsin}\,(H/R)$, so that $r = R\,\cos(\lambda)$ and $H = R\,\sin(\lambda)$, with $t_p = \varphi$ and $t_q = \lambda$:

$$\vec{\mathbf{p}}(t_p) = \begin{pmatrix} r\,\cos(t_p) \\ r\,\sin(t_p) \\ H \end{pmatrix} = \begin{pmatrix} R\,\cos(\lambda)\,\cos(\varphi) \\ R\,\cos(\lambda)\,\sin(\varphi) \\ R\,\sin(\lambda) \end{pmatrix} = \vec{\mathbf{q}}(t_q).$$

At that intersection point,

$$\begin{aligned}
\vec{\mathbf{p}}'(t_p) &= \begin{pmatrix} -r\,\sin(t_p) \\ r\,\cos(t_p) \\ 0 \end{pmatrix} = \begin{pmatrix} -R\,\cos(\lambda)\,\sin(\varphi) \\ R\,\cos(\lambda)\,\cos(\varphi) \\ 0 \end{pmatrix} \\
\vec{\mathbf{q}}'(t_q) &= \begin{pmatrix} -R\,\cos(\varphi)\sin(t_q) \\ -R\,\sin(\varphi)\sin(t_q) \\ R\,\cos(t_q) \end{pmatrix} = \begin{pmatrix} -R\,\cos(\varphi)\sin(\lambda) \\ -R\,\sin(\varphi)\sin(\lambda) \\ R\,\cos(\lambda) \end{pmatrix} \\
\cos(\theta) &= \frac{\langle \vec{\mathbf{p}}'(t_p), \vec{\mathbf{q}}'(t_q) \rangle}{\|\vec{\mathbf{p}}'(t_p)\| \cdot \|\vec{\mathbf{q}}'(t_q)\|} = \frac{0}{R \cdot R} = 0 \\
\theta &= \text{Arccos}\,(0) = \frac{\pi}{2}.
\end{aligned}$$

Thus, the lines tangent to the parallel $\vec{\mathbf{p}}$ and to the meridian $\vec{\mathbf{q}}$ at their point of intersection make a right angle, which means that they are perpendicular to each other. (See **Figures 8** and **9**.) □

Definition 2. With the meridian at longitude $-\varphi$ parametrized as just described by $\vec{\mathbf{q}}$, and at the point at longitude $-\varphi$ and latitude λ, the direction of *due North* is the vector $\vec{\mathbf{q}}'(\lambda)$ tangent to the meridian at that point, as already observed in **Example 7**.

Also, the *azimuth* of a curve $\vec{\mathbf{f}}$ at the point $\vec{\mathbf{f}}(t)$ is the angle α measured clockwise from due North to the vector $\vec{\mathbf{f}}'(t)$ tangent to the curve at $\vec{\mathbf{f}}(t)$. □

Exercises

33. Calculate the angle between the vectors $\vec{\mathbf{u}} = (4, 1, 8)$ and $\vec{\mathbf{v}} = (0, 1, 1)$.

34. Develop a method to parametrize the great circle $\vec{\mathbf{g}}$ that passes through the point $\vec{\mathbf{A}}$ with a specified tangent vector $\vec{\mathbf{W}}$ on the sphere $\mathcal{S}^2(0, R)$. Such a method would determine the path of an object moving along a great circle, given only its initial position $\vec{\mathbf{A}}$ and its initial speed $\vec{\mathbf{W}}$.

35. For the great circle through $\vec{\mathbf{A}}$ and $\vec{\mathbf{B}}$, develop a method to determine the time t at which $\vec{\mathbf{g}}(t) = \vec{\mathbf{B}}$. Such a method would yield the time necessary for an object, such as a ship or an aircraft, to travel from $\vec{\mathbf{A}}$ to $\vec{\mathbf{B}}$ along the shortest path on the surface of the planet.

36. If a navigator at a position $\vec{\mathbf{g}}(t)$ travels along a great circle $\vec{\mathbf{g}} : [0, 2\pi] \to \mathcal{S}^2 \subset \mathbb{R}^3$ from $\vec{\mathbf{A}} \in \mathcal{S}^2$ to $\vec{\mathbf{B}} \in \mathcal{S}^2$, to what azimuth $\alpha(t)$ should the navigator direct the craft at time t?

37. Consider a navigator traveling along the great circle from Antananarivo (Madagascar), at latitude $\lambda = -18°55'$ (South) and at longitude $-\varphi = -47°31'$ (East), to Bombay (India), at latitude $\lambda = 18°56'$ (North) and at longitude $-\varphi = -72°51'$ (East). Compute the azimuths (directions) α and β in which the navigator leaves Antananarivo (α) and arrives in Bombay (β). In other words, in what direction α, measured clockwise from due North, should the Navigator leave Antananarivo? Then compute the distance from Antananarivo to Bombay along that great circle, assuming for the Earth a radius of about 6378 kilometers [Snyder 1987b, 11].

38. Write a differential equation, involving $\vec{\mathbf{f}}$ and $\vec{\mathbf{f}}'$, to impose on a curve $\vec{\mathbf{f}}$ the condition that it have the same constant azimuth α at all its points. Such a curve, called a *loxodrome,* traces the trajectory of a navigator that constantly travels in the same direction measured from due North. On a sphere, sketch such a curve that starts at $\vec{\mathbf{f}}(0) = (1, 0, 0)$ and maintains a constant azimuth $\alpha = \pi/4$.

3. Differentials of Functions of Several Variables

3.1 Differentials of Real-Valued Functions of Several Variables

Recall from elementary multivariable calculus that a function $f : \mathcal{D} \subseteq \mathbb{R}^3 \to \mathbb{R}$ has for each variable a separate partial derivative, denoted here by $D_1 f$,

$D_2 f$, and $D_3 f$, and defined by

$$\begin{aligned} D_1 f(x, y, z) &= \lim_{h \to 0} \tfrac{f(x+h,y,z)-f(x,y,z)}{h}, \\ D_2 f(x, y, z) &= \lim_{k \to 0} \tfrac{f(x,y+k,z)-f(x,y,z)}{k}, \\ D_3 f(x, y, z) &= \lim_{\ell \to 0} \tfrac{f(x,y,z+\ell)-f(x,y,z)}{\ell}, \end{aligned}$$

provided that the limits exist. (Some calculus texts denote the partial derivatives by variables, for instance, $D_x f(x, y, z)$, which may seem convenient in simple situations. Yet after evaluation or substitution, such notation may prove ambiguous or cumbersome to decipher. For example, if $f(x, y, z) \equiv x \cdot y^2 \cdot z^3$, then $D_2 f(x, y, z) = 2xyz^3$ and hence $D_2 f(u, v, w) = 2uvz^3$, but what is $D_y(u \cdot v^2 \cdot w^3)$?) Thus,

$D_1 f(x, y, z)$ is the derivative of the function $x \mapsto f(x, y, z)$,
$D_2 f(x, y, z)$ is the derivative of the function $y \mapsto f(x, y, z)$,
$D_3 f(x, y, z)$ is the derivative of the function $z \mapsto f(x, y, z)$.

Consequently, many theorems about the derivative also hold for the partial derivatives, for instance, the following version of the mean value theorem.

Theorem 2. Mean Value Theorem for Partial Derivatives. *If $f : \mathcal{D} \subseteq \mathbb{R}^3 \to \mathbb{R}$ is continuous on an open set $\mathcal{D}$, if $[a, b] \times \{(y, z)\} \subseteq \mathcal{D}$, and if $D_1 f(x, y, z)$ exists for each $x \in]a, b[$, then there is a $c \in]a, b[$ for which*

$$f(b, y, z) - f(a, y, z) = (b - a)\, D_1 f(c, y, z).$$

Similar statements also hold for the other variables.

Proof: Apply the mean value theorem for derivatives to the function $x \mapsto f(x, y, z)$. □

The partial derivatives correspond to derivatives only in the directions of the axes of coordinates. Yet the mean value theorem for partial derivatives allows *continuous* partial derivatives to express the "directional derivatives" of functions of several variables in all directions, as explained by the following definition and theorem.

Definition 3. For each vector $\vec{\mathbf{u}} = (u, v, w)$ of unit length, and for each function $f : \mathcal{D} \subseteq \mathbb{R}^3 \to \mathbb{R}$, the **directional derivative** of f at $\vec{\mathbf{x}} = (x, y, z)$ in the direction $\vec{\mathbf{u}}$ is denoted by $D_{\vec{\mathbf{u}}} f(\vec{\mathbf{x}})$ and defined by

$$D_{\vec{\mathbf{u}}} f(\vec{\mathbf{x}}) = \lim_{t \to 0} \frac{f(x + tu, y + tv, z + tw) - f(x, y, z)}{t} = \lim_{t \to 0} \frac{f(\vec{\mathbf{x}} + t\vec{\mathbf{u}}) - f(\vec{\mathbf{x}})}{t}$$

provided that the limits exist. □

Though the foregoing definition appears to require a separate limit for each direction, the following theorem shows that if the partial derivatives are continuous, then they form the coordinates of the directional derivative. To this end, denote the dot product by $\vec{\mathbf{u}} \cdot \vec{\mathbf{v}}$ or by $\langle \vec{\mathbf{u}}, \vec{\mathbf{v}} \rangle$.

Theorem 3. *If all the partial derivatives of* $f : \mathcal{D} \subseteq \mathbb{R}^3 \to \mathbb{R}$ *are continuous, then*

$$D_{(\vec{\mathbf{u}})} f(\vec{\mathbf{x}}) = D_1 f(\vec{\mathbf{x}}) u + D_2 f(\vec{\mathbf{x}}) v + D_3 f(\vec{\mathbf{x}}) w.$$

Proof: Subtract and add two terms, and apply the mean value theorem for partial derivatives:

$$\begin{aligned}
&f(x+tu, y+tv, z+tw) - f(x,y,z)\\
&= f(x+tu, y+tv, z+tw) - f(x, y+tv, z+tw) + f(x, y+tv, z+tw)\\
&\quad - f(x,y,z+tw) + f(x,y,z+tw) - f(x,y,z)\\
&= tu\, D_1 f(c_1, y+tv, z+tw) + tv\, D_2 f(x, c_2, z+tw) + tw\, D_3 f(x,y,c_3)
\end{aligned}$$

with c_1 between x and $x+tu$, with c_2 between y and $y+tv$, and with c_3 between z and $z+tw$. Hence,

$$\begin{aligned}
&D_{\vec{\mathbf{u}}} f(x,y,z)\\
&= \lim_{t\to 0} \frac{f(x+tu, y+tv, z+tw) - f(x,y,z)}{t}\\
&= \lim_{t\to 0} \frac{tu\, D_1 f(c_1, y+tv, z+tw) + tv\, D_2 f(x, c_2, z+tw) + tw\, D_3 f(x,y,c_3)}{t}\\
&= \lim_{t\to 0} \{u\, D_1 f(c_1, y+tv, z+tw) + v\, D_2 f(x, c_2, z+tw) + w\, D_3 f(x,y,c_3)\}\\
&= u\, D_1 f(x,y,z) + v\, D_2 f(x,y,z) + w\, D_3 f(x,y,z)\\
&= \langle\, [D_1 f(\vec{\mathbf{x}}), D_2 f(\vec{\mathbf{x}}), D_3 f(\vec{\mathbf{x}})],\ (u,v,w)\rangle.
\end{aligned}$$

□

Definition 4. The *gradient* of f at $\vec{\mathbf{x}} = (x,y,z)$ is the vector denoted by $\vec{\nabla}(f)(\vec{\mathbf{x}})$ or $\vec{\text{grad}}(f)(\vec{\mathbf{x}})$ and defined by $\vec{\nabla}(f)(\vec{\mathbf{x}}) = (D_1 f(\vec{\mathbf{x}}), D_2 f(\vec{\mathbf{x}}), D_3 f(\vec{\mathbf{x}}))$ provided that the partial derivatives exist. □

The usefulness of the gradient is that it can express and calculate the directional derivatives in all directions. As for functions on the real line, the definition of the concept of limit means that for every positive number ε, a positive number δ exists such that

$$\left| \frac{f(\vec{\mathbf{x}} + t\,\vec{\mathbf{u}}) - f(\vec{\mathbf{x}})}{t} - \langle \vec{\nabla}(f)(\vec{\mathbf{x}}), \vec{\mathbf{u}}\rangle \right| < \varepsilon$$

for every nonzero real number t such that $0 < |t| < \delta$, or, equivalently,

$$\left| f(\vec{\mathbf{x}} + t\,\vec{\mathbf{u}}) - \left\{ f(\vec{\mathbf{x}}) + t\,\langle \vec{\nabla}(f)(\vec{\mathbf{x}}), \vec{\mathbf{u}}\rangle \right\} \right| < \varepsilon |t|, \tag{4}$$

for every real number t such that $|t| < \delta$. For nonzero vectors $\vec{\mathrm{h}}$ of any length $t > 0$, setting $\vec{\mathrm{u}} = \vec{\mathrm{h}}/\|\vec{\mathrm{h}}\| = \vec{\mathrm{h}}/t$ produces a unit vector, and the linearity of the dot product gives

$$t\,\langle\vec{\nabla}(f)(\vec{\mathrm{x}}),\vec{\mathrm{u}}\rangle = \langle\vec{\nabla}(f)(\vec{\mathrm{x}}),t\vec{\mathrm{u}}\rangle = \langle\vec{\nabla}(f)(\vec{\mathrm{x}}),\vec{\mathrm{h}}\rangle.$$

Substituting this expression into **(4)** for the gradient gives

$$\left|f(\vec{\mathrm{x}}+\vec{\mathrm{h}}) - \left\{f(\vec{\mathrm{x}}) + \langle\vec{\nabla}(f)(\vec{\mathrm{x}}),\vec{\mathrm{h}}\rangle\right\}\right|$$

$$= \left|f(\vec{\mathrm{x}}+t\,\vec{\mathrm{u}}) - \left\{f(\vec{\mathrm{x}}) + t\,\langle\vec{\nabla}(f)(\vec{\mathrm{x}}),\vec{\mathrm{u}}\rangle\right\}\right| < \varepsilon|t| = \varepsilon\|\vec{\mathrm{h}}\|,$$

$$\frac{\left|f(\vec{\mathrm{x}}+\vec{\mathrm{h}}) - \left\{f(\vec{\mathrm{x}}) + \langle\vec{\nabla}(f)(\vec{\mathrm{x}}),\vec{\mathrm{h}}\rangle\right\}\right|}{\|\vec{\mathrm{h}}\|} < \varepsilon,$$

for every vector $\vec{\mathrm{h}} \in \mathbb{R}^3$ such that $0 < \|\vec{\mathrm{h}}\| < \delta$, whence

$$\lim_{\vec{\mathrm{h}}\to\vec{0}} \frac{\left|f(\vec{\mathrm{x}}+\vec{\mathrm{h}}) - \left\{f(\vec{\mathrm{x}}) + \langle\vec{\nabla}(f)(\vec{\mathrm{x}}),\vec{\mathrm{h}}\rangle\right\}\right|}{\|\vec{\mathrm{h}}\|} = 0. \tag{5}$$

Inequalities similar to those used for functions on the real line reveal that the function $T_{f,\vec{\mathrm{x}}} : \mathbb{R}^2 \to \mathbb{R}$ defined by

$$T_{f,\vec{\mathrm{x}}}(\vec{\mathrm{x}}+\vec{\mathrm{h}}) = f(\vec{\mathrm{x}}) + \langle\vec{\nabla}(f)(\vec{\mathrm{x}}),\vec{\mathrm{h}}\rangle$$

forms a plane, denoted by $T_{f,\vec{\mathrm{x}}}$, which in each direction remains closer to f near $\vec{\mathrm{x}}$ than does any other plane. In other words, for each vector $\vec{\mathrm{G}}$ and each direction $\vec{\mathrm{u}}$, there is a positive number δ such that

$$\left|f(\vec{\mathrm{x}}+t\,\vec{\mathrm{u}}) - \left\{f(\vec{\mathrm{x}}) + t\,\langle\vec{\mathrm{G}},\vec{\mathrm{u}}\rangle\right\}\right| \geq \left|f(\vec{\mathrm{x}}+t\,\vec{\mathrm{u}}) - \left\{f(\vec{\mathrm{x}}) + t\,\langle\vec{\nabla}(f)(\vec{\mathrm{x}}),\vec{\mathrm{u}}\rangle\right\}\right|$$

for every t with $|t| < \delta$. Geometrically, therefore, the plane $T_{f,\vec{\mathrm{x}}}$ is called the plane *tangent* to f at $\vec{\mathrm{x}}$.

Analytically, the inequalities just obtained mean that among all vectors $\vec{\mathrm{G}} \in \mathbb{R}^3$, the vector $\vec{\nabla}(f)(\vec{\mathrm{x}})$ corresponds to the closest approximation of f near $\vec{\mathrm{x}}$ by affine functions. Moreover, $\vec{\nabla}(f)(\vec{\mathrm{x}})$ is the only vector for which

$$\lim_{\vec{\mathrm{h}}\to\vec{0}} \frac{\left|f(\vec{\mathrm{x}}+\vec{\mathrm{h}}) - \left\{f(\vec{\mathrm{x}}) + \langle\vec{\nabla}(f)(\vec{\mathrm{x}}),\vec{\mathrm{h}}\rangle\right\}\right|}{\|\vec{\mathrm{h}}\|} = 0.$$

Therefore, by similarlity with functions on the real line, the *differential* of f at $\vec{\mathrm{x}}$ is the function $D_{f,\vec{\mathrm{x}}} : \mathbb{R}^3 \to \mathbb{R}$ defined by

$$D_{f,\vec{\mathrm{x}}}(\vec{\mathrm{h}}) = \langle\vec{\nabla}(f)(\vec{\mathrm{x}}),\vec{\mathrm{h}}\rangle.$$

Exercises

39. This exercise shows how limit **(5)** for the differential and gradient distinguishes the tangent plane from all other planes. Consider a function $f : \mathcal{D} \subseteq \mathbb{R}^3 \to \mathbb{R}$ with continuous partial derivatives on an open set $\mathcal{D} \subseteq \mathbb{R}^3$. Also consider an affine function $T : \mathbb{R}^3 \to \mathbb{R}$ defined by a vector $\vec{\mathrm{G}} \neq \vec{\nabla}(f)(\vec{\mathrm{x}})$ and by the formula $T(\vec{\mathrm{x}}+\vec{\mathrm{h}}) = f(\vec{\mathrm{x}}) + \langle \vec{\mathrm{G}}, \vec{\mathrm{h}} \rangle$. Prove that

$$\begin{aligned}\lim_{\vec{\mathrm{h}}\to\vec{0}} \frac{\left|f(\vec{\mathrm{x}}+\vec{\mathrm{h}})-\left\{f(\vec{\mathrm{x}})+\langle\vec{\nabla}(f)(\vec{\mathrm{x}}),\vec{\mathrm{h}}\rangle\right\}\right|}{\|\vec{\mathrm{h}}\|} &= 0,\\ \lim_{\vec{\mathrm{h}}\to\vec{0}} \frac{\left|f(\vec{\mathrm{x}}+\vec{\mathrm{h}})-\left\{f(\vec{\mathrm{x}})+\langle\vec{\mathrm{G}},\vec{\mathrm{h}}\rangle\right\}\right|}{\|\vec{\mathrm{h}}\|} &\neq 0.\end{aligned}$$

40. This exercise shows that without the quotient by $\|\vec{\mathrm{h}}\|$, limit **(5)** for the differential and gradient would *not* distinguish the tangent plane from all the other planes. Consider a function $f : \mathcal{D} \subseteq \mathbb{R}^3 \to \mathbb{R}$ with continuous partial derivatives on an open set $\mathcal{D} \subseteq \mathbb{R}^3$. Also consider an affine function $T : \mathbb{R}^3 \to \mathbb{R}$ defined by any vector $\vec{\mathrm{G}} \in \mathbb{R}^3$ and by the formula $T(\vec{\mathrm{x}}+\vec{\mathrm{h}}) = f(\vec{\mathrm{x}}) + \langle \vec{\mathrm{G}}, \vec{\mathrm{h}} \rangle$. Prove that

$$\lim_{\vec{\mathrm{h}}\to\vec{0}} \left|f(\vec{\mathrm{x}}+\vec{\mathrm{h}}) - \left\{f(\vec{\mathrm{x}}) + \langle\vec{\mathrm{G}},\vec{\mathrm{h}}\rangle\right\}\right| = 0.$$

41. Recall that the perimeter, or length, $\ell(a, b)$ of an ellipse with principal semi-axes of lengths a and b is expressed by the formula

$$\ell(a, b) = \int_0^{2\pi} \sqrt{a^2 \sin^2(\theta) + b^2 \cos^2(\theta)}\, d\theta.$$

Also recall that the *area* bounded by the same ellipse is $A(a, b) = \pi ab$. Among all ellipses that bound an area π, what ellipses have the shortest perimeter?

42. Verify that the directional derivative equals the dot product of the direction vector with the gradient: $D_{\vec{\mathrm{u}}} f_{\vec{\mathrm{x}}} = \langle \vec{\nabla}(f)(\vec{\mathrm{x}}), \vec{\mathrm{u}} \rangle$.

43. Consider a function $f : \mathcal{D} \subseteq \mathbb{R}^3 \to \mathbb{R}$ with continuous partial derivatives on an open set $\mathcal{D} \subseteq \mathbb{R}^3$. Also consider an affine function $T : \mathbb{R}^3 \to \mathbb{R}$ defined by a vector $\vec{\mathrm{G}} \neq \vec{\nabla}(f)(\vec{\mathrm{x}})$ and by the formula $T(\vec{\mathrm{x}}+\vec{\mathrm{h}}) = f(\vec{\mathrm{x}}) + \langle \vec{\mathrm{G}}, \vec{\mathrm{h}} \rangle$. Prove that in each direction and in a neighborhood of $\vec{\mathrm{x}}$, the tangent plane $T_{f,\vec{\mathrm{x}}}$ remains closer to f than does the other affine function T.

44. Provide an example of a function f and a point $\vec{\mathrm{x}}$ such that there is no neighborhood of $\vec{\mathrm{x}}$ in which the tangent plane remains closer to f than any other plane does in the entire neighborhood. Such an example demonstrates the necessity of mentioning the direction before the neighborhood in the preceding exercise.

3.2 Level Sets of Functions of Several Variables

The differential and the gradient provide qualitative information about sets where the values of a function remain constant, because such sets, called "isopleths" or "level sets," remain perpendicular to the gradient.

Definition 5. A *level set* of a function $f : \mathbb{R}^3 \to \mathbb{R}$ is a subset $S \subseteq \mathbb{R}^3$ for which a constant $K \in \mathbb{R}$ exists such that $f(\vec{\mathbf{x}}) = K$ for every $\vec{\mathbf{x}} \in S$; thus, $S \subseteq f^{-1}(K)$. □

Example 12. Consider the function $f : \mathbb{R}^3 \to \mathbb{R}$ defined by $f(x, y, z) = x^2 + y^2 + z^2$. Then

$$\begin{aligned} \vec{\nabla}(f)(\vec{\mathbf{x}}) &= (D_1 f(\vec{\mathbf{x}}) + D_2 f(\vec{\mathbf{x}}) + D_3 f(\vec{\mathbf{x}})) \\ &= (2x, 2y, 2z) = 2(x, y, z) = 2\vec{\mathbf{x}}. \end{aligned}$$

For each positive number R, the level set where $f(x, y, z) = R^2$ consists of the sphere $\mathcal{S}^2(0, R)$ with radius R centered at the origin. At a point $\vec{\mathbf{x}} \in \mathcal{S}^2(0, R)$, the gradient $\vec{\nabla}(f)(\vec{\mathbf{x}}) = 2\vec{\mathbf{x}}$ points in the same direction as the radius vector $\vec{\mathbf{x}}$, and, consequently, points perpendicularly to the surface of the sphere. □

The following special case of the chain rule for differentials generalizes the preceding example.

Proposition 2. *If the curve* $\vec{\mathbf{g}} :]a, b[\to \mathbb{R}^3$ *has a derivative* $\vec{\mathbf{g}}'(t)$ *at* $t \in]a, b[$, *and if* $f : \mathcal{D} \subseteq \mathbb{R}^3 \to \mathbb{R}$ *has continuous partial derivatives at* $\vec{\mathbf{g}}(t)$ *and* $\vec{\mathbf{g}}(t) \in \mathcal{D}$, *then the differential at* t *of the composite function* $f \circ \vec{\mathbf{g}}$ *equals the product of the differential of* f *at* $\vec{\mathbf{g}}$ *and the differential of* $\vec{\mathbf{g}}$ *at* t:

$$(f \circ \vec{\mathbf{g}})'(t) = \langle \vec{\nabla}(f)(\vec{\mathbf{g}}(t)), \vec{\mathbf{g}}(t) \rangle.$$

Proof: First, apply the mean value theorem for derivatives to each coordinate of $\vec{\mathbf{g}}$:

$$\begin{aligned} g_1(t+h) - g_1(t) &= h\, g_1'(c_1), \\ g_2(t+h) - g_2(t) &= h\, g_2'(c_2), \\ g_3(t+h) - g_3(t) &= h\, g_3'(c_3), \end{aligned}$$

with c_1, c_2, and c_3 between t and $t + h$. Second, substitute the expressions just obtained, and apply the mean value theorem for partial derivatives:

$$\begin{aligned} &(f \circ \vec{\mathbf{g}})(t+h) - (f \circ \vec{\mathbf{g}})(t) \\ &\quad = [f(g_1(t+h), g_2(t+h), g_3(t+h))] - [f(g_1(t), g_2(t), g_3(t))] \\ &\quad = [f(g_1(t) + h g_1'(c_1), g_2(t) + h g_2'(c_2), g_3(t) + h g_3'(c_3))] \\ &\qquad\quad - [f(g_1(t), g_2(t), g_3(t))] \end{aligned}$$

$$\begin{aligned}
&= [f(g_1(t) + hg_1'(c_1), g_2(t) + hg_2'(c_2), g_3(t) + hg_3'(c_3))] \\
&\quad - [f(g_1(t), g_2(t) + hg_2'(c_2), g_3(t) + hg_3'(c_3))] \\
&\quad + [f(g_1(t), g_2(t) + hg_2'(c_2), g_3(t) + hg_3'(c_3))] \\
&\quad - [f(g_1(t), g_2(t), g_3(t) + hg_3'(c_3))] \\
&\quad + [f(g_1(t), g_2(t), g_3(t) + hg_3'(c_3))] - [f(g_1(t), g_2(t), g_3(t))] \\
&= hg_1'(c_1) \cdot D_1 f(g_1(t) + s_1 hg_1'(c_1), g_2(t) + hg_2'(c_2), g_3(t) + hg_3'(c_3)) \\
&\quad + hg_2'(c_2) \cdot D_2 f(g_1(t), g_2(t) + s_2 hg_2'(c_2), g_3(t) + hg_3'(c_3)) \\
&\quad + hg_3'(c_3) \cdot D_3 f(g_1(t), g_2(t), g_3(t) + s_3 hg_3'(c_3))
\end{aligned}$$

with $0 < s_1, s_2, s_3 < 1$. Dividing throughout by h and taking the limits as h tends to zero then causes c_1, c_2, and c_3 to tend to t and yields

$$\begin{aligned}
(f \circ \vec{g})'(t) &= \lim_{h \to 0} \frac{(f \circ \vec{g})(t+h) - (f \circ \vec{g})(t)}{h} \\
&= \lim_{h \to 0} (g_1'(c_1) \cdot D_1 f(g_1(t) + s_1 hg_1'(c_1), g_2(t) + hg_2'(c_2), g_3(t) + hg_3'(c_3)) \\
&\quad + g_2'(c_2) \cdot D_2 f(g_1(t), g_2(t) + s_2 hg_2'(c_2), g_3(t) + hg_3'(c_3)) \\
&\quad + g_3'(c_3) \cdot D_3 f(g_1(t), g_2(t), g_3(t) + s_3 hg_3'(c_3))) \\
&= g_1'(t) \cdot D_1 f(g_1(t), g_2(t), g_3(t)) + g_2'(t) \cdot D_2 f(g_1(t), g_2(t), g_3(t)) \\
&\quad + g_3'(t) \cdot D_3 f(g_1(t), g_2(t), g_3(t)) \\
&= \langle \vec{g}'(t), \vec{\nabla}(f)(\vec{g}(t)) \rangle.
\end{aligned}$$

□

Remark 5. The foregoing proof also demonstrates the need for symbols for summations over indices. Indeed, the foregoing proof includes every argument three times, one each for x, y, and z, and is therefore three times as long as a proof that would include only one argument for an indexed variable x_i. □

Theorem 4. *If $f : \mathcal{D} \subseteq \mathbb{R}^3 \to \mathbb{R}$ has continuous partial derivatives, if a constant $K \in \mathbb{R}$ exists for which $f(\vec{x}) = K$ for every $\vec{x}$ in an open level set $\mathcal{L} \subseteq \mathcal{D}$, and if $\vec{g} :]a, b[\to \mathbb{R}^3$ is a curve such that $\vec{g}(t) \in \mathcal{L}$ for every $t \in]a, b[$, then $\vec{g}'(t) \perp \vec{\nabla}(f)(\vec{g}(t))$.*

Proof: Apply the chain rule:

$$\langle \vec{\nabla}(f)(\vec{g}(t)), \vec{g}'(t) \rangle = (f \circ \vec{g})'(t) = K' = 0.$$

□

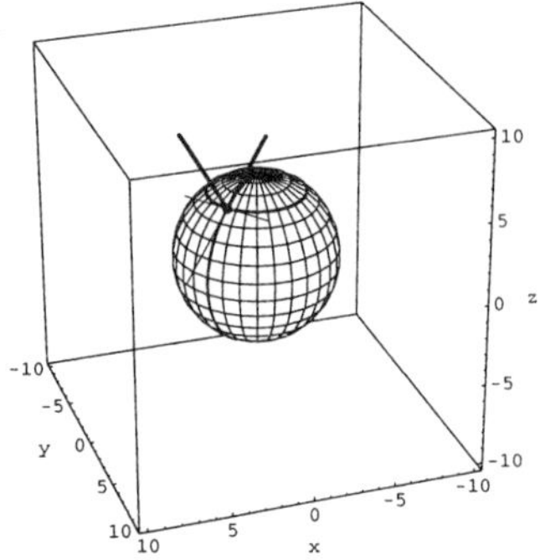

Figure 10. Because the parallels and meridians lie on the sphere, which is a level set of a function f, the lines tangent to parallels or meridians are perpendicular to the gradient of that function f.

Example 13. The derivatives in **Examples 5** and **7** illustrate how at each point $\vec{\mathbf{x}}$ on the sphere $S^2(0,R)$, the tangent vector to the meridian through $\vec{\mathbf{x}}$ and the tangent vector to the parallel through $\vec{\mathbf{x}}$ are perpendicular to the radius vector $\vec{\mathbf{x}}$, and hence to the gradient $2\vec{\mathbf{x}}$ of $f(x,y,z) = x^2 + y^2 + z^2$ at $\vec{\mathbf{x}}$. For the parallels from **Example 5,**

$$\vec{\mathbf{p}}(t) = \begin{pmatrix} r\cos(t) \\ r\sin(t) \\ H \end{pmatrix}, \quad \vec{\mathbf{p}}'(t) = \begin{pmatrix} -r\sin(t) \\ r\cos(t) \\ 0 \end{pmatrix},$$

$$\begin{aligned} \langle \vec{\mathbf{p}}'(t), \vec{\nabla}(f)(\vec{\mathbf{p}}(t)) \rangle &= \langle \vec{\mathbf{p}}'(t), 2\vec{\mathbf{p}}(t) \rangle \\ &= \begin{pmatrix} r\cos(t) & r\sin(t) & H \end{pmatrix} 2 \begin{pmatrix} -r\sin(t) \\ r\cos(t) \\ 0 \end{pmatrix} = 0. \end{aligned}$$

Similarly, for the meridians from **Example 7,** denoted here by $\vec{\mathbf{q}}$,

$$\vec{\mathbf{q}}(t) = \begin{pmatrix} R\cos(\varphi)\cos(t) \\ R\sin(\varphi)\cos(t) \\ R\sin(t) \end{pmatrix}, \quad \vec{\mathbf{q}}'(t) = \begin{pmatrix} -R\cos(\varphi)\sin(t) \\ -R\sin(\varphi)\sin(t) \\ R\cos(t) \end{pmatrix},$$

$$\begin{aligned} \langle \vec{\mathbf{q}}'(t), \vec{\nabla}(f)(\vec{\mathbf{q}}(t)) \rangle &= \langle \vec{\mathbf{q}}'(t), 2\vec{\mathbf{q}}(t) \rangle \\ &= \left\langle \begin{pmatrix} R\cos(\varphi)\cos(t) \\ R\sin(\varphi)\cos(t) \\ R\sin(t) \end{pmatrix}, 2 \begin{pmatrix} -R\cos(\varphi)\sin(t) \\ -R\sin(\varphi)\sin(t) \\ R\cos(t) \end{pmatrix} \right\rangle \\ &= 0. \end{aligned}$$

□

Theorem 5. *Consider a smooth curve* $\vec{\mathbf{g}} : \mathcal{I} \subseteq \mathbb{R} \to \mathbb{R}^n$ *defined on an interval* $\mathcal{I}$. *Also, consider a differentiable function* $f : \mathcal{D} \subseteq \mathbb{R}^n \to \mathbb{R}$ *defined on an open set* $\mathcal{D}$. *Moreover, let* $t \in \mathcal{I}$ *denote a number such that* $\vec{\mathbf{g}}(t) \in \mathcal{D}$ *and* $\|\vec{\mathbf{g}}'(t)\| = 1$. *Then the direction of* $\vec{\mathbf{g}}'(t)$ *that maximizes the rate of change* $(f \circ \vec{\mathbf{g}})'(t)$ *is the direction of* $\vec{\nabla}(f)$ *at the point* $\vec{\mathbf{g}}(t)$.

Proof: Apply the chain rule and the Cauchy–Schwartz inequality:

$$\begin{aligned} (f \circ \vec{\mathbf{g}})'(t) &= \langle \vec{\nabla}(f)(\vec{\mathbf{g}}(t)), \vec{\mathbf{g}}'(t) \rangle \\ &\leq \left| \langle \vec{\nabla}(f)(\vec{\mathbf{g}}(t)), \vec{\mathbf{g}}'(t) \rangle \right| \\ &\leq \left\| \vec{\nabla}(f)(\vec{\mathbf{g}}(t)) \right\| \cdot \|\vec{\mathbf{g}}'(t)\|, \end{aligned}$$

with equality if, but only if, $\vec{\nabla}(f)(\vec{\mathbf{g}}(t))$ and $\vec{\mathbf{g}}'(t)$ point in parallel directions, which means that one of the two vectors equals a scalar multiple of the other vector. □

Application 7. On a topographic geographical map, the level curves represent horizontal curves, along each of which the altitude of the terrain remains constant. Thus, if the plane $\mathbb{R}^2$ represents a map, and if the function $f : \mathbb{R}^2 \to \mathbb{R}$ gives the altitude $f(x, y)$ of a point located by (x, y) on the map $\mathbb{R}^2$, then the level curves are the level sets of f. Moreover, at each point (x, y) on the map, the level curves lie perpendicularly to the gradient $\vec{\nabla}(f)(x, y)$. Furthermore, the image of a curve $\vec{\mathbf{g}}$, which may represent a path, rises the steepest if, but only if, the direction $\vec{\mathbf{g}}'(t)$ tangent to the path points in the direction of the gradient $\vec{\nabla}(f)(\vec{\mathbf{g}}(t))$ at the point $\vec{\mathbf{g}}(t)$. In contrast, if $\vec{\mathbf{g}}'(t)$ points perpendicularly to $\vec{\nabla}(f)(\vec{\mathbf{g}}(t))$, then the path is horizontal, parallel to a level curve, at $\vec{\mathbf{g}}(t)$. For exercises with level curves in elementary multivariable calculus, see Curjel [1990]. □

Exercises

45. Prove that the tangents to each great circle indeed lie perpendicularly to the radius vector from the center of the Earth to the point of tangency.

46. State and prove a Mean Value Theorem for Directional Derivatives.

4. Sample Examination Problems

Each of the following problems illustrates a type of problem on maps that may appear in examinations or homework at the level of multivariable calculus.

1. This problem introduces the *Date Line,* which is the most popular dating service in the world. The Date Line consists of the half of the meridian at longitude $\varphi = 180°$, without the other half at longitude $\varphi = 0°$. As the Earth rotates counterclockwise seen from the North Pole, each new calendar day begins as the Date Line passes through midnight, opposite to the sun on the dark side of the Earth. At midnight on the Date Line, the entire Earth is on the same calendar day. Then the region of the Earth immediately to the West of the Date Line, with positive longitudes close to $180°$, lies one calendar day ahead of the region immediately to the East of the Date Line, with negative longitudes close to $-180°$.

   Identify the entire region that lies one day ahead of the remaining portion of the Earth.

2. Does each parallel cross every meridian at an angle that remains constant along that parallel? In other words, is each parallel a rhumb line?

3. Calculate the distance along the great circle through Canberra (Australia), at longitude $149°8'$ (East) and latitude $-35°15'$ (South), and Vancouver (Canada), at longitude $-123°7'$ (West) and latitude $49°16'$ (North). Assume for the Earth a radius of about 6378 kilometers. Also calculate the azimuths of that great circle in Canberra and in Vancouver.

4. The present problem demonstrates an application of mathematics to supplement geographical maps that may lack certain desirable characteristics, for instance, the USDA's Forest Service's map *Colville National Forest* (Washington State).

   On 19 June 1994, I stood by the remains of a fire-lookout at altitude 6782 ft on top of Columbia Mountain, at longitude $-118°28'48''$ (West) and latitude $48°37'13''$ (North). Under the azimuth $\alpha = 64°$, measured with an accuracy of perhaps $\pm 2°$ with a pocket compass, I saw a mountain that looked like 7308-ft high Abercombie, about 50 miles away at longitude $-117°26'02''$ (West) and latitude $48°55'40''$ (North). The map confirmed that no obstacle should block the line of sight, but Columbia Mountain and Abercombie Mountain appeared on the two opposite sides of the map, which did not allow for a graphic verification of the azimuth.

   Under what azimuth does one see Abercombie Mountain from Columbia Mountain? Given the accuracy of the compass measurements, could the mountain seen in the azimuth $64° \pm 2°$ have been Abercombie Mountain? In what azimuth does one see Columbia from Abercombie?

5. Does each great circle cross every meridian at an angle that remains constant along that great circle? In other words, is each great circle a rhumb line?

6. For each spatial curve $\vec{\mathbf{f}} : \mathcal{D} \subseteq \mathbb{R} \to \mathbb{R}^3$, for each $t \in \mathcal{D}$, and for each vector $\vec{\mathbf{w}} \in \mathbb{R}^3$, with $\vec{\mathbf{w}} \neq \vec{\mathbf{f}}'(t)$, consider the parametric line $\vec{\mathbf{L}} : \mathbb{R} \to \mathbb{R}^3$ defined by $\vec{\mathbf{L}}(t+h) = \vec{\mathbf{f}}(t) + h\,\vec{\mathbf{w}}$. Prove that a positive number δ exists such that

$$\left\|\vec{\mathbf{f}}(t+h) - \vec{\mathbf{L}}(t+h)\right\| > \left\|\vec{\mathbf{f}}(t+h) - \vec{\mathbf{T}}_{\vec{\mathbf{f}},t}(t+h)\right\|$$

for every h such that $0 < |h| < \delta$.

5. Solutions to the Odd-Numbered Exercises

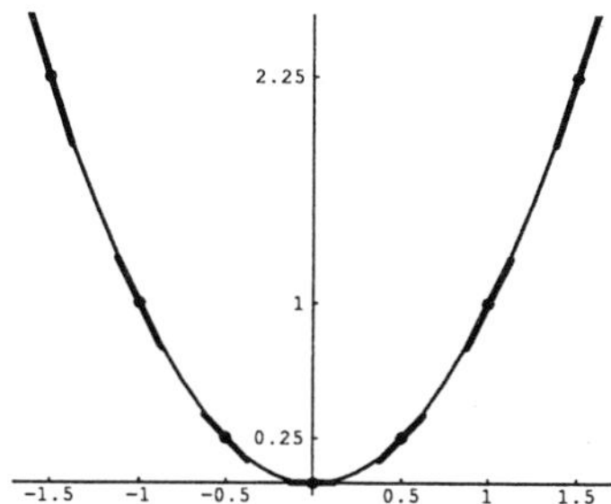

Figure 11. Solution to **Exercise 1.**

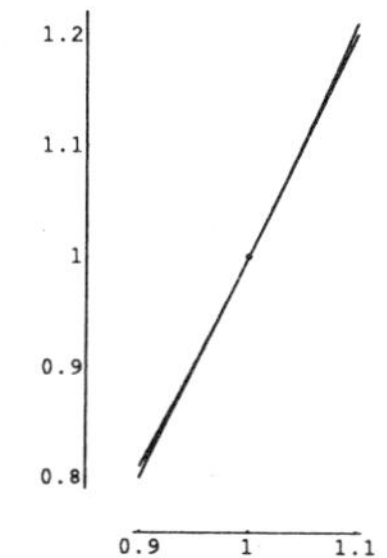

Figure 12. Solution to **Exercise 1.**

1. See **Figure 11.**

3. See **Figure 12.**

5. For the tangent line, apply the definition of the derivative:

$$\begin{aligned}
\lim_{h\to 0} \frac{f(x+h) - T_{f,x}(x+h)}{h} &= \lim_{h\to 0} \frac{f(x+h) - \{f(x) + h\,f'(x)\}}{h} \\
&= \lim_{h\to 0} \left\{ \frac{f(x+h) - f(x)}{h} - f'(x) \right\} \\
&= \left\{ \lim_{h\to 0} \frac{f(x+h) - f(x)}{h} \right\} - f'(x) \\
&= f'(x) - f'(x) = 0.
\end{aligned}$$

For every line through $(x, f(x))$ different from the tangent line,

$$\lim_{h\to 0}\frac{f(x+h)-L(x+h)}{h}$$

$$\begin{aligned}
&= \lim_{h\to 0}\frac{f(x+h)-T_{f,x}(x+h)+T_{f,x}(x+h)-L(x+h)}{h}\\
&= \lim_{h\to 0}\frac{f(x+h)-T_{f,x}(x+h)}{h}+\lim_{h\to 0}\frac{T_{f,x}(x+h)-L(x+h)}{h}\\
&= 0+\lim_{h\to 0}\frac{\{f(x)+h\,f'(x)\}-\{f(x)+h\,m\}}{h}\\
&= \lim_{h\to 0}\frac{h\,f'(x)-h\,m}{h}=f'(x)-m\neq 0.
\end{aligned}$$

7. Apply the chain rule: $g'(x)=r\,f'(r^{-1}x)\,r^{-1}=f'(r^{-1}x)$.

9. Let $g(x)=f(x)-f(0)$. Then $g(0)=f(0)-f(0)=0$. Moreover,

$$\begin{aligned}
|g(w)-g(v)| &= |\{f(w)-f(0)\}-\{f(v)-f(0)\}|\\
&= |f(w)-f(v)|=|w-v|.
\end{aligned}$$

Thus, g fixes the origin and preserves distances. The particular choice $v=0$ gives $|g(w)|=|g(w)-0|=|g(w)-g(0)|=|w-0|=|w|$, whence $[g(w)]^2=w^2$. Applying the equation just obtained to $w=x-z$ gives

$$\begin{aligned}
x^2-2g(x)g(z)+z^2 &= [g(x)]^2-2g(x)g(z)+[g(z)]^2=\{g(x)-g(z)\}^2\\
&= |g(x)-g(z)|^2=|x-z|^2=(x-z)^2\\
&= x^2-2xz+z^2,
\end{aligned}$$

and canceling similar terms produces $g(x)g(z)=xz$. The particular choice $z=1$ then yields $g(x)g(1)=x$, but the previous equation $[g(x)]^2=x^2$ has the consequence that $g(x)=\pm x$, so that $g(1)=\pm 1$. It follows that either $g(x)=x$ for every x, or $g(x)=-x$ for every x. Returning to f and setting $c=f(0)$ confirms that $f(x)=g(x)+f(0)=g(x)+c$, which gives either $f(x)=c+x$ for every x or $f(x)=c-x$ for every x.

11. By definition of limits in the plane, for each $\varepsilon>0$, some $\delta>0$ exists such that

$$\left\|\frac{\vec{\mathrm{f}}(t+h)-\vec{\mathrm{f}}(t)}{h}-\vec{\mathrm{f}}'(t)\right\|<\varepsilon$$

for every h such that $0 < |h| < \delta$. Also, by definition of the Euclidean norm,

$$\left\| \frac{\vec{\mathbf{f}}(t+h) - \vec{\mathbf{f}}(t)}{h} - \vec{\mathbf{f}}'(t) \right\|$$

$$= \sqrt{\left\{[f_1(t+h) - f_1(t)]/h - [\vec{\mathbf{f}}']_1(t)\right\}^2 + \left\{[f_2(t+h) - f_2(t)]/h - [\vec{\mathbf{f}}']_2(t)\right\}^2}.$$

Consequently,

$$\left| \, [f_1(t+h) - f_1(t)]/h - [\vec{\mathbf{f}}']_1(t) \, \right|$$

$$\leq \sqrt{\left\{[f_1(t+h) - f_1(t)]/h - [\vec{\mathbf{f}}']_1(t)\right\}^2 + \left\{[f_2(t+h) - f_2(t)]/h - [\vec{\mathbf{f}}']_2(t)\right\}^2}$$

$$= \left\| \frac{\vec{\mathbf{f}}(t+h) - \vec{\mathbf{f}}(t)}{h} - \vec{\mathbf{f}}'(t) \right\| < \varepsilon$$

for every h such that $0 < |h| < \delta$, which means that

$$[\vec{\mathbf{f}}']_1(t) = \lim_{h \to 0} [f_1(t+h) - f_1(t)]/h = f_1'(t).$$

Similarly,

$$[\vec{\mathbf{f}}']_2(t) = \lim_{h \to 0} [f_2(t+h) - f_2(t)]/h = f_2'(t).$$

13. For every nonzero h sufficiently close to zero,

$$\left\| \vec{\mathbf{f}}(t+h) - \vec{\mathbf{L}}(t+h) \right\| > \left\| \vec{\mathbf{f}}(t+h) - \vec{\mathbf{T}}_{\vec{\mathbf{f}},t}(t+h) \right\|.$$

To verify that this inequality holds, write the parametric "point-direction" equation of $\vec{\mathbf{L}}$ in the form $\vec{\mathbf{L}}(t+h) = \vec{\mathbf{f}}(t) + h \cdot \vec{\mathbf{m}}$, and compare

$$\left\| \vec{\mathbf{f}}(t+h) - \vec{\mathbf{L}}(t+h) \right\| \qquad \text{to} \qquad \left\| \vec{\mathbf{f}}(t+h) - \vec{\mathbf{T}}_{\vec{\mathbf{f}},x}(t+h) \right\|$$

in the open interval $]t-\delta, t+\delta[$, with δ corresponding to any ε such that $0 < \varepsilon < \|\vec{\mathbf{m}} - \vec{\mathbf{f}}'(t)\|/2$:

$$\left\| \vec{\mathbf{f}}(t+h) - \vec{\mathbf{L}}(t+h) \right\|$$

$$= \left\| \vec{\mathbf{f}}(t+h) - \{\vec{\mathbf{f}}(t) + h \cdot \vec{\mathbf{m}}\} \right\|$$

$$= \left\| \vec{\mathbf{f}}(t+h) - \{\vec{\mathbf{f}}(t) + h\,\vec{\mathbf{f}}'(t) - h\,\vec{\mathbf{f}}'(t) + h \cdot \vec{\mathbf{m}}\} \right\|$$

$$
\begin{aligned}
&= \left\|\vec{\mathbf{f}}(t+h) - \{\vec{\mathbf{f}}(t) + h\,\vec{\mathbf{f}}'(t)\} - \{h\cdot\vec{\mathrm{m}} - h\,\vec{\mathbf{f}}'(t)\}\right\| \\
&= \left\|\vec{\mathbf{f}}(t+h) - \{\vec{\mathbf{f}}(t) + h\,\vec{\mathbf{f}}'(t)\} - h\cdot\{\vec{\mathrm{m}} - \vec{\mathbf{f}}'(t)\}\right\| \\
&\geq \left|\left\|\vec{\mathbf{f}}(t+h) - \{\vec{\mathbf{f}}(t) + h\,\vec{\mathbf{f}}'(t)\}\right\| - \left\|h\cdot\{\vec{\mathrm{m}} - \vec{\mathbf{f}}'(t)\}\right\|\right| \\
&= \left|\left\|h\cdot\{\vec{\mathrm{m}} - \vec{\mathbf{f}}'(t)\}\right\| - \left\|\vec{\mathbf{f}}(t+h) - \{\vec{\mathbf{f}}(t) + h\,\vec{\mathbf{f}}'(t)\}\right\|\right| \\
&= \left\|h\cdot\{\vec{\mathrm{m}} - \vec{\mathbf{f}}'(t)\}\right\| - \left\|\vec{\mathbf{f}}(t+h) - \{\vec{\mathbf{f}}(t) + h\,\vec{\mathbf{f}}'(t)\}\right\| \\
&\geq |h|\cdot\left\|\vec{\mathrm{m}} - \vec{\mathbf{f}}'(t)\right\| - \varepsilon|h| \\
&> 2\varepsilon|h| - \varepsilon|h| = \varepsilon|h| \\
&> \left\|\vec{\mathbf{f}}(t+h) - \vec{\mathbf{T}}_{\vec{\mathbf{f}},t}(t+h)\right\|.
\end{aligned}
$$

Therefore, $\vec{\mathbf{T}}_{\vec{\mathbf{f}},t}$ remains closer to $\vec{\mathbf{f}}$ than $\vec{\mathbf{L}}$ does. For this reason, $\vec{\mathbf{T}}_{\vec{\mathbf{f}},t}$ is called the line **tangent** to $\vec{\mathbf{f}}$ at $(t, \vec{\mathbf{f}}(t))$. Thus, the derivative $\vec{\mathbf{f}}'(t)$ represents the slope of the tangent line $\vec{\mathbf{T}}_{\vec{\mathbf{f}},t}$.

15. The points $v = 0$ and $w = 1$ lie at a distance $|w - v| = |1 - 0| = 1$ apart, but their images $\vec{\mathrm{s}}(v) = \vec{\mathrm{s}}(0) = (0,1)$ and $\vec{\mathrm{s}}(w) = \vec{\mathrm{s}}(1) = (1,0)$ lie at a distance $\pi/2$ measured along the arc of the unit circle from $(0,1)$ to $(1,0)$.

17. The hypotheses on $\vec{\mathbf{f}}$ imply $\int_v^w \|\vec{\mathbf{f}}'(t)\|\,dt = w - v$ for all $v, w \in \mathbb{R}$ such that $v < w$. By contraposition, assume that some $t \in \mathbb{R}$ exists for which $\|\vec{\mathbf{f}}'(t)\| \neq 1$, for instance, $\|\vec{\mathbf{f}}'(t)\| > 1$. Let

$$\varepsilon = \frac{\|\vec{\mathbf{f}}'(t)\| - 1}{2}.$$

By continuity of $\|\vec{\mathbf{f}}'(t)\|$, a positive δ exists such that $\left|\|\vec{\mathbf{f}}'(s)\| - \|\vec{\mathbf{f}}'(t)\|\right| < \varepsilon$, for every s such that $0 \leq |s - t| < \delta$. Consequently, a rearrangement gives

$$
\begin{aligned}
\|\vec{\mathbf{f}}'(t)\| - \varepsilon &< \|\vec{\mathbf{f}}'(s)\| < \|\vec{\mathbf{f}}'(t)\| + \varepsilon \\
\|\vec{\mathbf{f}}'(t)\| - \frac{\|\vec{\mathbf{f}}'(t)\| - 1}{2} &< \|\vec{\mathbf{f}}'(s)\| < \|\vec{\mathbf{f}}'(t)\| + \frac{\|\vec{\mathbf{f}}'(t)\| - 1}{2} \\
\frac{\|\vec{\mathbf{f}}'(t)\| + 1}{2} &< \|\vec{\mathbf{f}}'(s)\| < \|\vec{\mathbf{f}}'(t)\| + \frac{\|\vec{\mathbf{f}}'(t)\| - 1}{2}
\end{aligned}
$$

whence

$$1 < \frac{\|\vec{\mathbf{f}}'(t)\| + 1}{2} < \|\vec{\mathbf{f}}'(s)\|$$

for every s such that $0 \le |s-t| < \delta$. Letting $v = t - \delta/2$ and $w = t + \delta/2$ then yields

$$\int_v^w \|\vec{\mathbf{f}}'(t)\|\,dt > \int_v^w 1\,dt = (w-v),$$

which would contradict the hypotheses on $\vec{\mathbf{f}}$.

19. From $z = (1-t^2)/(1+t^2)$ follows $z(1+t^2) = 1-t^2$, whence $t^2(1+z) = 1-z$, and hence $t^2 = (1-z)/(1+z)$. From $x = 2t/(1+t^2)$ follows $x(1+t^2)/2 = t$, where the foregoing result gives $1+t^2 = 1 + [(1-z)/(1+z)] = 2/(1+z)$. Consequently, $t = x(1+t^2)/2 = x/(1+z)$.

21. By definition of a parabola, and with D denoting the point on L closest to P, the points on the parabola lie at equal distance from the focus and from the directrix: $\overline{PF} = \overline{PD}$. The formula for the derivative then shows that $T \perp [DF]$, whence T bisects the angle DPF. See Ayoub [1991].

23. As demonstrated in **Example 11**, at an intersection of the parallel $\vec{\mathbf{p}}$ at latitude $\text{Arcsin}\,(H/R)$ and the meridian $\vec{\mathbf{q}}$ at longitude $-\varphi$,

$$\vec{\mathbf{p}}(u) = \begin{pmatrix} r\,\cos(u) \\ r\,\sin(u) \\ H \end{pmatrix} = \vec{\mathbf{q}}(v) = \begin{pmatrix} R\,\cos(\varphi)\cos(v) \\ R\,\sin(\varphi)\cos(v) \\ R\,\sin(v) \end{pmatrix}.$$

Equating the third coordinates gives $H = R\,\sin(v)$, whence $\sin(v) = H/R$ and $\cos(v) = \pm\sqrt{1-(H/R)^2} = \pm[\sqrt{R^2-H^2}]/R = \pm r/R$. Hence, with $R\,\cos(v) = r$, equating the first and second coordinates respectively gives $u = \varphi$. Substituting the values of u and v just obtained confirms that *two* intersections exist: $(r\,\cos(\varphi), r\,\sin(\varphi), H)$ and $(-r\,\cos(\varphi), -r\,\sin(\varphi), H)$.

25. Let $\varepsilon = |1-b| \cdot \left\|\vec{\mathbf{f}}'(t)\right\|/2$; with the corresponding δ, for each h such that $|h| < \delta$,

$$\begin{aligned}
&\left\|\vec{\mathbf{f}}(t+h) - \vec{\mathbf{L}}(t+h)\right\| \\
&= \left\|\vec{\mathbf{f}}(t+h) - \vec{\mathbf{T}}_{\vec{\mathbf{f}},t}(t+h) + \vec{\mathbf{T}}_{\vec{\mathbf{f}},t}(t+h) - \vec{\mathbf{L}}(t+h)\right\| \\
&\ge \left|\left\|\vec{\mathbf{f}}(t+h) - \vec{\mathbf{T}}_{\vec{\mathbf{f}},t}(t+h)\right\| - \left\|\vec{\mathbf{T}}_{\vec{\mathbf{f}},t}(t+h) - \vec{\mathbf{L}}(t+h)\right\|\right| \\
&= \left|\left\|\vec{\mathbf{f}}(t+h) - \left\{\vec{\mathbf{f}}(t) + h\,\vec{\mathbf{f}}'(t)\right\}\right\|\right. \\
&\qquad \left. - \left\|\left\{\vec{\mathbf{f}}(t) + h\,\vec{\mathbf{f}}'(t)\right\} - \left\{\vec{\mathbf{f}}(t) + bh\,\vec{\mathbf{f}}'(t)\right\}\right\|\right| \\
&= \left|\left\|\vec{\mathbf{f}}(t+h) - \left\{\vec{\mathbf{f}}(t) + h\,\vec{\mathbf{f}}'(t)\right\}\right\| - \left\|h\left\{\vec{\mathbf{f}}'(t) - b\,\vec{\mathbf{f}}'(t)\right\}\right\|\right|
\end{aligned}$$

$$
\begin{aligned}
&\geq \ |h| \cdot |1-b| \cdot \left\|\vec{\mathbf{f}}'(t)\right\| - |h|\varepsilon \\
&\geq \ 2|h|\varepsilon - |h|\varepsilon = |h|\varepsilon \\
&> \ \left\|\vec{\mathbf{f}}(t+h) - \vec{\mathbf{T}}_{\vec{\mathbf{f}},t}(t+h)\right\|.
\end{aligned}
$$

27. *Step 1.* $\vec{\mathbf{N}} = \vec{\mathbf{B}} \times \vec{\mathbf{A}} = (6,7,6) \times (2,6,9) = (27,-42,22)$.
Step 2. $\vec{\mathbf{G}} = \vec{\mathbf{A}} \times \vec{\mathbf{N}} = (2,6,9) \times (27,-42,22) = (510,199,-246)$.
Step 3. $\|\vec{\mathbf{G}}\| = \sqrt{360{,}217}$ and $\vec{\mathbf{V}} = \vec{\mathbf{G}}/\|\vec{\mathbf{G}}\| = (510,199,-246)/\sqrt{360{,}217}$.
Step 4. $\vec{\mathbf{U}} = \vec{\mathbf{A}}/\|\vec{\mathbf{A}}\| = \vec{\mathbf{A}}/R = (2,6,9)/11$.
Consequently, the following function $\vec{\mathbf{g}}$ traces the great circle through $(2,6,9)$ and $(6,7,6)$:

$$
\begin{aligned}
\vec{\mathbf{g}}(t) &= 11\cos(t)\,\frac{1}{11}\begin{pmatrix}2\\6\\9\end{pmatrix} + 11\sin(t)\,\frac{1}{\sqrt{360{,}217}}\begin{pmatrix}510\\199\\-246\end{pmatrix} \\
&= \begin{pmatrix}2\cos(t) + 11 \times 510\sin(t)/\sqrt{360{,}217}\\ 6\cos(t) + 11 \times 199\sin(t)/\sqrt{360{,}217}\\ 9\cos(t) - 11 \times 246\sin(t)/\sqrt{360{,}217}\end{pmatrix}.
\end{aligned}
$$

29. From the orthogonality of $\vec{\mathbf{U}}$ and $\vec{\mathbf{V}}$, and from the formula $\vec{\mathbf{g}}'(t) = -R\sin(t)\vec{\mathbf{U}} + R\cos(t)\vec{\mathbf{V}}$, it follows that $\|\vec{\mathbf{g}}'(t)\| = R$. Consequently, the function $t \mapsto \vec{\mathbf{g}}(t \cdot v/R)$ parametrizes the same great circle as $\vec{\mathbf{g}}$, but with the desired speed $R \cdot v/R = v$, obtained by the chain rule.

31. The Terminator T consists of the intersection of the spherical Earth E and the plane P passing through the center O of the Earth and lying perpendicularly to the direction $\vec{\ell}$ of the light rays from the sun, with the direction $\vec{\ell}$ oriented from the sun toward the Earth. Indeed, for each point X on the Earth, consider the half-line L_X from X in the direction $-\vec{\ell}$, toward the sun. If the sun and the point X lie on the same side of the plane P, then L_X does not encounter any other point on Earth, so that X lies in the sunlight. In contrast, if the sun and the point X lie on opposite sides of the plane P, then L_X encounters a symmetric point on Earth, so that X lies in the dark.

The foregoing outline of a geometric proof also admits an algebraic counterpart, in which the plane P has equation $\langle\vec{\mathbf{x}},\vec{\ell}\rangle = 0$. Algebra then confirms that the sun and the hemisphere in the sunlight (day) lie in the region where $\langle\vec{\mathbf{x}},\vec{\ell}\rangle < 0$, while the hemisphere in the dark (night) lies in the region where $\langle\vec{\mathbf{x}},\vec{\ell}\rangle > 0$.

33. $$\langle \vec{\mathbf{u}}, \vec{\mathbf{v}} \rangle = \langle (4,1,8), (0,1,1) \rangle = 9; \qquad \|\vec{\mathbf{u}}\| = \|(4,1,8)\| = \sqrt{81} = 9;$$
$$\|\vec{\mathbf{v}}\| = \|(0,1,1)\| = \sqrt{2}$$
$$\cos(\theta) = \frac{\langle \vec{\mathbf{u}}, \vec{\mathbf{v}} \rangle}{\|\vec{\mathbf{u}}\| \cdot \|\vec{\mathbf{v}}\|} = \frac{9}{9 \times \sqrt{2}} = \frac{1}{\sqrt{2}}; \qquad \theta = \mathrm{Arccos}\left(\frac{1}{\sqrt{2}}\right) = \frac{\pi}{4}.$$

35. While moving from $\vec{\mathbf{A}}$ to $\vec{\mathbf{B}}$ along the great circle $\vec{\mathbf{g}}$, an object traces an arc of that circle subtending the angle θ, measured in radians, between the two vectors $\vec{\mathbf{A}}$ and $\vec{\mathbf{B}}$; hence,

$$\theta = \mathrm{Arccos}\left(\frac{\langle \vec{\mathbf{A}}, \vec{\mathbf{B}} \rangle}{\|\vec{\mathbf{A}}\| \cdot \|\vec{\mathbf{B}}\|}\right).$$

Consequently, the arc of great circle from $\vec{\mathbf{A}}$ to $\vec{\mathbf{B}}$ has length $L = R\theta$. Traveling at speed v, according to the formula $t \mapsto \vec{\mathbf{g}}(t \cdot v/R)$, an object starting from $\vec{\mathbf{A}}$ at time 0 reaches $\vec{\mathbf{B}}$ at time $t = L/v = R\theta/v$.

37. For Antananarivo (Madagascar),

$$\begin{aligned}
\vec{\mathbf{A}} &= \begin{pmatrix} \cos(\varphi)\cos(\lambda) \\ \sin(\varphi)\cos(\lambda) \\ \sin(\lambda) \end{pmatrix} = \begin{pmatrix} \cos(47°31')\cos(-18°55') \\ \sin(47°31')\cos(-18°55') \\ \sin(-18°55') \end{pmatrix} \\
&\approx \begin{pmatrix} (0.675\,375\,713\,752)(0.945\,991\,095\,044) \\ (0.737\,473\,826\,840)(0.945\,991\,095\,044) \\ -0.324\,192\,609\,566 \end{pmatrix} \\
&\approx \begin{pmatrix} 0.638\,899\,411\,018 \\ 0.697\,643\,673\,019 \\ -0.324\,192\,609\,566 \end{pmatrix}.
\end{aligned}$$

For Bombay (India),

$$\begin{aligned}
\vec{\mathbf{B}} &= \begin{pmatrix} \cos(\varphi)\cos(\lambda) \\ \sin(\varphi)\cos(\lambda) \\ \sin(\lambda) \end{pmatrix} = \begin{pmatrix} \cos(72°51')\cos(18°56') \\ \sin(72°51')\cos(18°56') \\ \sin(18°56') \end{pmatrix} \\
&\approx \begin{pmatrix} (0.294\,874\,299\,918)(0.945\,896\,751\,215) \\ (0.955\,536\,052\,301)(0.945\,896\,751\,215) \\ 0.324\,467\,773\,500 \end{pmatrix} \\
&\approx \begin{pmatrix} 0.278\,920\,642\,309 \\ 0.903\,838\,447\,540 \\ 0.324\,467\,773\,500 \end{pmatrix}.
\end{aligned}$$

Hence,

$$\vec{\mathbf{N}} = \vec{\mathbf{B}} \times \vec{\mathbf{A}} = \begin{pmatrix} -0.519\,380\,634\,215 \\ 0.297\,726\,280\,275 \\ -0.382\,874\,630\,407 \end{pmatrix},$$

$$\vec{\mathbf{G}} = \vec{\mathbf{A}} \times \vec{\mathbf{N}} = \begin{pmatrix} -0.170\,589\,403\,724 \\ 0.412\,997\,739\,025 \\ 0.552\,559\,758\,461 \end{pmatrix},$$

$$\|\vec{\mathbf{G}}\| = 0.710\,626\,599\,399, \qquad \vec{\mathbf{V}} = \frac{\vec{\mathbf{G}}}{\|\vec{\mathbf{G}}\|} = \begin{pmatrix} -0.240\,054\,909\,102 \\ 0.581\,174\,050\,304 \\ 0.777\,566\,951\,378 \end{pmatrix}.$$

Consequently,

$$\begin{aligned} \vec{\mathbf{g}}(t) &= \cos(t)\vec{\mathbf{A}} + \sin(t)\vec{\mathbf{V}} \\ &\approx \cos(t)\begin{pmatrix} 0.638\,899\,411\,018 \\ 0.697\,643\,673\,019 \\ -0.324\,192\,609\,566 \end{pmatrix} + \sin(t)\begin{pmatrix} -0.240\,054\,909\,102 \\ 0.581\,174\,050\,304 \\ 0.777\,566\,951\,378 \end{pmatrix} \\ &= \begin{pmatrix} 0.638\,899\,411\,018\cos(t) - 0.240\,054\,909\,102\sin(t) \\ 0.697\,643\,673\,019\cos(t) + 0.581\,174\,050\,304\sin(t) \\ -0.324\,192\,609\,566\cos(t) + 0.777\,566\,951\,378\sin(t) \end{pmatrix}. \end{aligned}$$

The arc of great circle from Antananarivo to Bombay corresponds to the angle

$$\begin{aligned} \theta &= \operatorname{Arccos}\left(\frac{\langle\vec{\mathbf{A}}, \vec{\mathbf{B}}\rangle}{\|\vec{\mathbf{A}}\| \cdot \|\vec{\mathbf{B}}\|}\right) = \operatorname{Arccos}\left(\frac{0.703\,569\,354\,239}{1 \cdot 1}\right) \\ &= 0.790\,388\,410\,019 \text{ radians} = 45°17'09''. \end{aligned}$$

With an average Earth's radius R approximately equal to 6378 kilometers, the distance from Antananarivo to Bombay becomes

$$R\theta = 6378 \times 0.790\,388\,410\,019 \approx 5041 \text{ km}.$$

In Antananarivo, the unit vector pointing toward due North is

$$\begin{aligned} \vec{\mathbf{N}}_{\vec{\mathbf{A}}} &= \begin{pmatrix} -\cos(\varphi)\sin(\lambda) \\ -\sin(\varphi)\sin(\lambda) \\ \cos(\lambda) \end{pmatrix} = \begin{pmatrix} -\cos(47°31')\sin(-18°55') \\ -\sin(47°31')\sin(-18°55') \\ \cos(-18°55') \end{pmatrix} \\ &= \begin{pmatrix} 0.218\,951\,815\,079 \\ 0.239\,083\,564\,410 \\ 0.945\,991\,095\,044 \end{pmatrix}. \end{aligned}$$

Consequently, with $\vec{g}'(0) = \vec{V}$ and $\|\vec{V}\| = 1$, the azimuth becomes

$$\begin{aligned}\alpha &= \text{Arccos}\left(\frac{\langle \vec{N}_{\vec{A}}, \vec{g}'(0)\rangle}{\|\vec{N}_{\vec{A}}\| \cdot \|\vec{g}'(0)\|}\right) = \text{Arccos}\,(0.821\,960\ldots)\\ &= 0.605\,952\ldots \text{ radian } = 34^\circ 43' 07''.\end{aligned}$$

In Bombay, the unit vector pointing toward due North is

$$\begin{aligned}\vec{N}_{\vec{B}} &= \begin{pmatrix} -\cos(\varphi)\sin(\lambda) \\ -\sin(\varphi)\sin(\lambda) \\ \cos(\lambda)\end{pmatrix} = \begin{pmatrix} -\cos(72^\circ 51')\sin(18^\circ 56') \\ -\sin(72^\circ 51')\sin(18^\circ 56') \\ \cos(18^\circ 56')\end{pmatrix}\\ &= \begin{pmatrix} -0.095\,677\,207\,556\,8 \\ -0.310\,040\,655\,389 \\ 0.945\,896\,751\,215\end{pmatrix}.\end{aligned}$$

Consequently, with $\vec{B} = \vec{g}(0.790\,388\,410\,019)$,

$$\vec{g}'(0.790\,388\,410\,019) = \cos(0.790\,388\,410\,019)\vec{V} - \sin(0.790\,388\,410\,019)\vec{A},$$

and $\|\vec{V}\| = 1 = \|\vec{A}\|$, the azimuth becomes

$$\begin{aligned}\beta &= \text{Arccos}\left(\frac{\langle \vec{N}_{\vec{B}}, \vec{g}'(0.790\,388\,410\,019)\rangle}{\|\vec{N}_{\vec{B}}\| \cdot \|\vec{g}'(0.790\,388\,410\,019)\|}\right)\\ &= \text{Arccos}\,(0.821\,921\ldots) = 0.606\,021\ldots \text{ radian } = 34^\circ 43' 21''.\end{aligned}$$

Other methods of solution exist, for instance, through spherical trigonometry [Vest and Benge 1994–1995], but the present method illustrates the use of the concepts introduced here.

39. The limit does not exist, because the value of the limit depends upon the direction of $\vec{h}$. Letting θ denote the angle between $\vec{h}$ and $\vec{G} - \vec{\nabla}(f)(\vec{x})$, so that $\langle \vec{G} - \vec{\nabla}(f)(\vec{x}), \vec{h}\rangle = \|\vec{G} - \vec{\nabla}(f)(\vec{x})\| \cdot \|\vec{h}\| \cdot \cos(\theta)$, gives

$$\lim_{\vec{h}\to\vec{0}} \frac{\left|f(\vec{x}+\vec{h}) - \left\{f(\vec{x}) + \langle \vec{G}, \vec{h}\rangle\right\}\right|}{\|\vec{h}\|}$$

$$\begin{aligned}&= \lim_{\vec{h}\to\vec{0}} \frac{\left|f(\vec{x}+\vec{h}) - \left\{f(\vec{x}) + \langle \vec{G} - \vec{\nabla}(f)(\vec{x}) + \vec{\nabla}(f)(\vec{x}), \vec{h}\rangle\right\}\right|}{\|\vec{h}\|}\\ &\geq \lim_{\vec{h}\to\vec{0}} \frac{\left|\langle \vec{G} - \vec{\nabla}(f)(\vec{x}), \vec{h}\rangle\right|}{\|\vec{h}\|} - \lim_{\vec{h}\to\vec{0}} \frac{\left|f(\vec{x}+\vec{h}) - f(\vec{x}) - \langle \vec{\nabla}(f)(\vec{x}), \vec{h}\rangle\right|}{\|\vec{h}\|}\end{aligned}$$

$$
\begin{aligned}
&= \lim_{\vec{\mathbf{h}} \to \vec{\mathbf{0}}} \left| \frac{\|\vec{\mathbf{G}} - \vec{\nabla}(f)(\vec{\mathbf{x}})\| \cdot \|\vec{\mathbf{h}}\| \cos(\theta)}{\|\vec{\mathbf{h}}\|} \right| - 0 \\
&= \|\vec{\mathbf{G}} - \vec{\nabla}(f)(\vec{\mathbf{x}})\| \cdot |\cos(\theta)| .
\end{aligned}
$$

41. The constraint that $A(a,b) = \pi$ means that $\pi ab = \pi$, whence $ab = 1$; consequently, $a \neq 0 \neq b$ and $b = 1/a$, which leaves only one variable. The problem thus reduces to finding the minimum of the function $f : \mathbb{R}_+^* =]0, \infty[\to \mathbb{R}$ defined by

$$f(a) = \ell(a, 1/a) = \int_0^{2\pi} \sqrt{a^2 \sin^2(\theta) + (1/a)^2 \cos^2(\theta)}\, d\theta.$$

Hence, setting the derivative equal to zero gives

$$0 = f'(a) = \int_0^{2\pi} \frac{a \sin^2(\theta) + (1/a)(-1/a^2)\cos^2(\theta)}{\sqrt{a^2 \sin^2(\theta) + (1/a)^2 \cos^2(\theta)}}\, d\theta.$$

However,

$$
\begin{aligned}
\int_0^{2\pi} \frac{\sin^2(\theta)}{\sqrt{\sin^2(\theta) + \cos^2(\theta)}}\, d\theta &= \int_0^{2\pi} \frac{\sin^2(\theta)}{1}\, d\theta = \int_0^{2\pi} \frac{1 - \cos(2\theta)}{2}\, d\theta \\
&= \left. \frac{\theta - \sin(2\theta)/2}{2} \right|_0^{2\pi} = \pi, \\
\int_0^{2\pi} \frac{\cos^2(\theta)}{\sqrt{\cos^2(\theta) + \sin^2(\theta)}}\, d\theta &= \int_0^{2\pi} \frac{\cos^2(\theta)}{1}\, d\theta = \int_0^{2\pi} \frac{1 + \cos(2\theta)}{2}\, d\theta \\
&= \left. \frac{\theta + \sin(2\theta)/2}{2} \right|_0^{2\pi} = \pi.
\end{aligned}
$$

Because both integrals equal each other, the foregoing results suggest that $f'(1) = 0$, and, indeed,

$$f'(1) = \left(1 - \frac{1}{1^3}\right) \cdot \pi = 0.$$

The second derivative test confirms that $a = 1$ corresponds to a *globally minimum* perimeter:

$$f''(a) =$$

$$\int_0^{2\pi} \frac{[a \sin^2(\theta) + (3/a^3)\cos^2(\theta)][a \sin^2(\theta) + (1/a^3)\cos^2(\theta)] - [a \sin^2(\theta) - (1/a^3)\cos^2(\theta)]^2}{[a^2 \sin^2(\theta) + (1/a)^2 \cos^2(\theta)]^{3/2}}\, d\theta > 0,$$

because

$$\begin{aligned}[a\sin^2(\theta) - (1/a^3)\cos^2(\theta)] &< [a\sin^2(\theta) + (1/a^3)\cos^2(\theta)] \\ &< [a\sin^2(\theta) + (3/a^3)\cos^2(\theta)]\end{aligned}$$

at every θ except $\pi/2$ and $3\pi/2$.

The solution just obtained, $a = 1$ and $b = 1/a = 1/1 = 1 = a$, means that *the ellipse with the least perimeter that encloses a specified area must be a circle.*

Other methods of solution exist, for instance, using Lagrange multipliers.

43. If $\vec{\mathrm{h}} \perp [\vec{\mathrm{G}} - \vec{\nabla}(f)(\vec{\mathrm{x}})]$, then both affine functions have the same value:

$$\left|f(\vec{\mathrm{x}}+\vec{\mathrm{h}}) - \left\{f(\vec{\mathrm{x}}) + \langle \vec{\mathrm{G}}, \vec{\mathrm{h}}\rangle\right\}\right|$$

$$\begin{aligned}
&= \left|f(\vec{\mathrm{x}}+\vec{\mathrm{h}}) - \left\{f(\vec{\mathrm{x}}) + \langle \vec{\mathrm{G}} - \vec{\nabla}(f)(\vec{\mathrm{x}}) + \vec{\nabla}(f)(\vec{\mathrm{x}}), \vec{\mathrm{h}}\rangle\right\}\right| \\
&= \left|f(\vec{\mathrm{x}}+\vec{\mathrm{h}}) - \left\{f(\vec{\mathrm{x}}) + \langle \vec{\nabla}(f)(\vec{\mathrm{x}}), \vec{\mathrm{h}}\rangle\right\} - \left\{\langle \vec{\mathrm{G}} - \vec{\nabla}(f)(\vec{\mathrm{x}}), \vec{\mathrm{h}}\rangle\right\}\right| \\
&= \left|f(\vec{\mathrm{x}}+\vec{\mathrm{h}}) - \left\{f(\vec{\mathrm{x}}) + \langle \vec{\nabla}(f)(\vec{\mathrm{x}}), \vec{\mathrm{h}}\rangle\right\}\right| - 0.
\end{aligned}$$

If $\vec{\mathrm{h}}$ does not lie perpendicularly to $\vec{\mathrm{G}} - \vec{\nabla}(f)(\vec{\mathrm{x}})$, let $\vec{\mathrm{u}} = \vec{\mathrm{h}}/\|\vec{\mathrm{h}}\|$ represent the unit vector in the direction of $\vec{\mathrm{h}}$. Adopt the notation for inequality **(4)**, and choose δ corresponding to $\varepsilon = \left|\langle \vec{\mathrm{G}} - \vec{\nabla}(f)(\vec{\mathrm{x}}), \vec{\mathrm{u}}\rangle\right|/2 > 0$. *Notice that ε and δ depend upon the angle between the direction $\vec{\mathrm{u}}$ and the vector $\vec{\mathrm{G}} - \vec{\nabla}(f)(\vec{\mathrm{x}})$.* For every $\vec{\mathrm{h}} \in \mathbb{R}^2$ such that $\|\vec{\mathrm{h}}\| < \delta$, inequality **(4)** holds, whence the reverse triangle inequality gives

$$\left|f(\vec{\mathrm{x}}+\vec{\mathrm{h}}) - \left\{f(\vec{\mathrm{x}}) + \langle \vec{\mathrm{G}}, \vec{\mathrm{h}}\rangle\right\}\right|$$

$$\begin{aligned}
&= \left|f(\vec{\mathrm{x}}+\vec{\mathrm{h}}) - \left\{f(\vec{\mathrm{x}}) + \langle \vec{\mathrm{G}} - \vec{\nabla}(f)(\vec{\mathrm{x}}) + \vec{\nabla}(f)(\vec{\mathrm{x}}), \vec{\mathrm{h}}\rangle\right\}\right| \\
&= \left|f(\vec{\mathrm{x}}+\vec{\mathrm{h}}) - \left\{f(\vec{\mathrm{x}}) + \langle \vec{\nabla}(f)(\vec{\mathrm{x}}), \vec{\mathrm{h}}\rangle\right\} - \left\{\langle \vec{\mathrm{G}} - \vec{\nabla}(f)(\vec{\mathrm{x}}), \vec{\mathrm{h}}\rangle\right\}\right| \\
&\geq \left|\langle \vec{\mathrm{G}} - \vec{\nabla}(f)(\vec{\mathrm{x}}), \vec{\mathrm{h}}\rangle\right| - \left|f(\vec{\mathrm{x}}+\vec{\mathrm{h}}) - \left\{f(\vec{\mathrm{x}}) + \langle \vec{\nabla}(f)(\vec{\mathrm{x}}), \vec{\mathrm{h}}\rangle\right\}\right| \\
&\geq 2\varepsilon\|\vec{\mathrm{h}}\| - \varepsilon\|\vec{\mathrm{h}}\| = \varepsilon\|\vec{\mathrm{h}}\| \\
&\geq \left|f(\vec{\mathrm{x}}+\vec{\mathrm{h}}) - \left\{f(\vec{\mathrm{x}}) + \langle \vec{\nabla}(f)(\vec{\mathrm{x}}), \vec{\mathrm{h}}\rangle\right\}\right|.
\end{aligned}$$

6. Solutions to the Sample Exam Problems

1. Call the half of the meridian opposite to the sun the *midnight line;* the other half of the same meridian then lies at noon directly "under" the sun. The "lune" to the west of the Date Line and to the east of the midnight line lies one calendar day ahead of the remaining portion of the Earth, the "lune" to the east of the Date Line and to the west of the midnight line.

2. Yes, each parallel *is* a rhumb line, because each parallel crosses each meridian at the same angle $\pi/2$ (or $-\pi/2$, depending upon the orientation). For a formal proof, verify that $\vec{\mathbf{p}}'(t_p)\cdot\vec{\mathbf{q}}'(t_q)=0$ at each point of intersection $\vec{\mathbf{p}}'(t_p)=\vec{\mathbf{x}}=\vec{\mathbf{q}}'(t_q)$.

3. For Canberra (Australia),

$$\begin{aligned}
\vec{\mathbf{A}} &= \begin{pmatrix}\cos(\varphi)\cos(\lambda)\\ \sin(\varphi)\cos(\lambda)\\ \sin(\lambda)\end{pmatrix} = \begin{pmatrix}\cos(149^\circ 8')\cos(-35^\circ 15')\\ \sin(149^\circ 8')\cos(-35^\circ 15')\\ \sin(-35^\circ 15')\end{pmatrix}\\
&\approx \begin{pmatrix}(-0.858\,363\,526\,682)(0.816\,641\,555\,162)\\ (0.513\,041\,963\,257)(0.816\,641\,555\,162)\\ -0.577\,145\,190\,037\end{pmatrix}\\
&\approx \begin{pmatrix}-0.700\,975\,325\,324\\ 0.418\,971\,386\,738\\ -0.577\,145\,190\,037\end{pmatrix}.
\end{aligned}$$

For Vancouver (Canada),

$$\begin{aligned}
\vec{\mathbf{B}} &= \begin{pmatrix}\cos(\varphi)\cos(\lambda)\\ \sin(\varphi)\cos(\lambda)\\ \sin(\lambda)\end{pmatrix} = \begin{pmatrix}\cos(-123^\circ 7')\cos(49^\circ 16')\\ \sin(-123^\circ 7')\cos(49^\circ 16')\\ \sin(49^\circ 16')\end{pmatrix}\\
&\approx \begin{pmatrix}(-0.546\,345\,620\,408)(0.652\,539\,358\,127)\\ (-0.837\,559\,826\,556)(0.652\,539\,358\,127)\\ 0.757\,754\,832\,446\end{pmatrix}\\
&\approx \begin{pmatrix}-0.356\,512\,020\,457\\ -0.546\,540\,751\,614\\ 0.757\,754\,832\,446\end{pmatrix}.
\end{aligned}$$

Hence,

$$\begin{aligned}
\vec{\mathbf{N}} &= \vec{\mathbf{B}}\times\vec{\mathbf{A}} = \begin{pmatrix}-0.002\,044\,227\,004\,1\\ -0.736\,926\,637\,987\\ -0.532\,479\,916\,765\end{pmatrix},\\
\vec{\mathbf{G}} &= \vec{\mathbf{A}}\times\vec{\mathbf{N}} = \begin{pmatrix}-0.648\,407\,513\,662\\ -0.372\,075\,467\,1\\ 0.517\,423\,862\,426\end{pmatrix},
\end{aligned}$$

$$\|\vec{\mathbf{G}}\| = 0.909\,175\,401\,338, \qquad \vec{\mathbf{V}} = \frac{\vec{\mathbf{G}}}{\|\vec{\mathbf{G}}\|} = \begin{pmatrix} -0.713\,181\,980\,844 \\ -0.409\,244\,977\,979 \\ 0.569\,113\,354\,435 \end{pmatrix}.$$

Consequently,

$$\begin{aligned}
\vec{\mathbf{g}}(t) &= \cos(t)\vec{\mathbf{A}} + \sin(t)\vec{\mathbf{V}} \\
&\approx \cos(t)\begin{pmatrix} -0.700\,975\,325\,324 \\ 0.418\,971\,386\,738 \\ -0.577\,145\,190\,037 \end{pmatrix} + \sin(t)\begin{pmatrix} -0.713\,181\,980\,844 \\ -0.409\,244\,977\,979 \\ 0.569\,113\,354\,435 \end{pmatrix} \\
&= \begin{pmatrix} -0.700\,975\,325\,324\cos(t) - 0.713\,181\,980\,844\sin(t) \\ 0.418\,971\,386\,738\cos(t) - 0.409\,244\,977\,979\sin(t) \\ -0.577\,145\,190\,037\cos(t) + 0.569\,113\,354\,435\sin(t) \end{pmatrix}.
\end{aligned}$$

The arc of great circle from Canberra to Vancouver corresponds to the angle

$$\begin{aligned}
\theta &= \operatorname{Arccos}\left(\frac{\langle\vec{\mathbf{A}}, \vec{\mathbf{B}}\rangle}{\|\vec{\mathbf{A}}\| \cdot \|\vec{\mathbf{B}}\|}\right) = \operatorname{Arccos}\left(\frac{-0.416\,413\,363\,864}{1 \cdot 1}\right) \\
&= 2.000\,293\,133\,13 \text{ radians} = 114°36'30''.
\end{aligned}$$

With an average Earth's radius R approximately equal to 6378 kilometers, the distance from Canberra to Vancouver becomes

$$R\theta = 6378 \times 2.000\,293\,133\,13 \approx 12{,}758 \text{ km}.$$

In Canberra, the unit vector pointing toward due North is

$$\begin{aligned}
\vec{\mathbf{N}}_{\vec{\mathbf{A}}} &= \begin{pmatrix} -\cos(\varphi)\sin(\lambda) \\ -\sin(\varphi)\sin(\lambda) \\ \cos(\lambda) \end{pmatrix} = \begin{pmatrix} -\cos(149°8')\sin(-35°15') \\ -\sin(149°8')\sin(-35°15') \\ \cos(-35°15') \end{pmatrix} \\
&= \begin{pmatrix} -0.495\,400\,380\,728 \\ 0.296\,099\,701\,381 \\ 0.816\,641\,555\,162 \end{pmatrix}.
\end{aligned}$$

Consequently, with $\vec{\mathbf{g}}'(0) = \vec{\mathbf{V}}$ and $\|\vec{\mathbf{V}}\| = 1$, the azimuth becomes

$$\begin{aligned}
\alpha &= \operatorname{Arccos}\left(\frac{\langle\vec{\mathbf{N}}_{\vec{\mathbf{A}}}, \vec{\mathbf{g}}'(0)\rangle}{\|\vec{\mathbf{N}}_{\vec{\mathbf{A}}}\| \cdot \|\vec{\mathbf{g}}'(0)\|}\right) = \operatorname{Arccos}(0.696\,894\ldots) \\
&= 0.799\,737\ldots \text{ radian } = 45°49'18''.
\end{aligned}$$

In Vancouver, the unit vector pointing toward due North is

$$\vec{\mathbf{N}}_{\vec{\mathbf{B}}} = \begin{pmatrix} -\cos(\varphi)\sin(\lambda) \\ -\sin(\varphi)\sin(\lambda) \\ \cos(\lambda) \end{pmatrix} = \begin{pmatrix} -\cos(-123°7')\sin(49°16') \\ -\sin(-123°7')\sin(49°16') \\ \cos(49°16') \end{pmatrix}$$

$$= \begin{pmatrix} 0.413\,996\,034\,050 \\ 0.634\,665\,006\,035 \\ 0.652\,539\,358\,127 \end{pmatrix}.$$

Consequently, with $\vec{\mathbf{B}} = \vec{\mathbf{g}}(2.000\,293\,133\,13)$ and

$$\begin{aligned} \vec{\mathbf{g}}'(2.000\,293\,133\,13) &= \cos(2.000\,293\,133\,13)\vec{\mathbf{V}} - \sin(2.000\,293\,133\,13)\vec{\mathbf{A}} \\ &= \begin{pmatrix} 0.934\,288\,030\,419 \\ -0.210\,503\,400\,761 \\ 0.287\,739\,803\,440 \end{pmatrix} \end{aligned}$$

and $\|\vec{\mathbf{V}}\| = 1 = \|\vec{\mathbf{A}}\|$, the azimuth becomes

$$\begin{aligned} \beta &= \operatorname{Arccos}\left(\frac{\langle \vec{\mathbf{N}}_{\vec{\mathbf{B}}}, \vec{\mathbf{g}}'(2.000\,293\,133\,13) \rangle}{\|\vec{\mathbf{N}}_{\vec{\mathbf{B}}}\| \cdot \|\vec{\mathbf{g}}'(2.000\,293\,133\,13)\|} \right) = \operatorname{Arccos}(0.440\,954\,\ldots\ldots) \\ &= 1.114\,135\ldots \text{ radian} = 63^\circ 50' 07''. \end{aligned}$$

4. For Columbia Mountain,

$$\begin{aligned} \vec{\mathbf{A}} &= \begin{pmatrix} \cos(\varphi)\cos(\lambda) \\ \sin(\varphi)\cos(\lambda) \\ \sin(\lambda) \end{pmatrix} = \begin{pmatrix} \cos(118^\circ 28' 48'')\cos(48^\circ 37' 13'') \\ \sin(118^\circ 28' 48'')\cos(48^\circ 37' 13'') \\ \sin(48^\circ 37' 13'') \end{pmatrix} \\ &\approx \begin{pmatrix} (-0.476\,851\,966\,153)(0.661\,046\,349\,113) \\ (0.878\,983\,618\,946)(0.661\,046\,349\,113) \\ 0.750\,345\,070\,167 \end{pmatrix} \approx \begin{pmatrix} -0.315\,221\,251\,293 \\ 0.581\,048\,912\,234 \\ 0.750\,345\,070\,167 \end{pmatrix}. \end{aligned}$$

For Abercombie Mountain,

$$\begin{aligned} \vec{\mathbf{B}} &= \begin{pmatrix} \cos(\varphi)\cos(\lambda) \\ \sin(\varphi)\cos(\lambda) \\ \sin(\lambda) \end{pmatrix} = \begin{pmatrix} \cos(117^\circ 26' 02'')\cos(48^\circ 55' 40'') \\ \sin(117^\circ 26' 02'')\cos(48^\circ 55' 40'') \\ \sin(48^\circ 55' 40'') \end{pmatrix} \\ &\approx \begin{pmatrix} (-0.460\,724\,822\,812)(0.657\,009\,830\,710) \\ (0.887\,543\,034\,250)(0.657\,009\,830\,710) \\ 0.753\,882\,008\,242 \end{pmatrix} \approx \begin{pmatrix} -0.302\,700\,737\,840 \\ 0.583\,124\,498\,680 \\ 0.753\,882\,008\,242 \end{pmatrix}. \end{aligned}$$

Hence,

$$\vec{\mathbf{N}} = \vec{\mathbf{B}} \times \vec{\mathbf{A}} = \begin{pmatrix} -4.977\,279\,636\,56 \times 10^{-4} \\ -1.050\,962\,359\,12 \times 10^{-2} \\ 7.929\,299\,679\,15 \times 10^{-3} \end{pmatrix}.$$

On Columbia Mountain, the unit vector pointing toward due North is

$$\vec{\mathbf{N}}_{\vec{\mathbf{A}}} = \begin{pmatrix} -\cos(\varphi)\sin(\lambda) \\ -\sin(\varphi)\sin(\lambda) \\ \cos(\lambda) \end{pmatrix} = \begin{pmatrix} -\cos(118^\circ 28' 48'')\sin(48^\circ 37' 13'') \\ -\sin(118^\circ 28' 48'')\sin(48^\circ 37' 13'') \\ \cos(48^\circ 37' 13'') \end{pmatrix}$$

$$\approx \begin{pmatrix} 0.357\,803\,522\,002 \\ -0.659\,541\,025\,234 \\ 0.661\,046\,349\,113 \end{pmatrix},$$

whereas the vector $\vec{g}'(0)$ tangent at Columbia Mountain to the great circle $\vec{g}$ through both mountains is

$$\vec{g}'(0) = \vec{V} = \begin{pmatrix} 0.948\,265\,950\,286 \\ 0.161\,370\,653\,814 \\ 0.273\,406\,656\,125 \end{pmatrix}.$$

Consequently, with $\vec{g}'(0) = \vec{V}$ and $\|\vec{V}\| = 1$, the azimuth becomes

$$\begin{aligned} \alpha &= \text{Arccos}\left(\frac{\langle \vec{N}_{\vec{A}}, \vec{g}'(0)\rangle}{\|\vec{N}_{\vec{A}}\| \cdot \|\vec{g}'(0)\|}\right) = \text{Arccos}\,(0.413\,597\ldots) \\ &= 1.144\,395\ldots \text{ radian } = 65°34'08'', \end{aligned}$$

which lies within the accuracy of the measured azimuth, $64° \pm 2°$. Thus, the mountain seen from Columbia Mountain in the azimuth $64° \pm 2°$ could have been Abercombie Mountain.

Because of the Earth's curvature, the azimuth on Columbia Mountain differs from that on Abercombie Mountain. On Abercombie Mountain, the unit vector pointing toward due North is

$$\begin{aligned} \vec{N}_{\vec{B}} &= \begin{pmatrix} -\cos(\varphi)\sin(\lambda) \\ -\sin(\varphi)\sin(\lambda) \\ \cos(\lambda) \end{pmatrix} = \begin{pmatrix} -\cos(117°26'02'')\sin(48°55'40'') \\ -\sin(117°26'02'')\sin(48°55'40'') \\ \cos(48°55'40'') \end{pmatrix} \\ &\approx \begin{pmatrix} 0.347\,332\,154\,668 \\ -0.669\,102\,725\,062 \\ 0.657\,009\,830\,710 \end{pmatrix}. \end{aligned}$$

Moreover, the vectors from the Earth's center to Columbia and Abercombie Mountain form an angle

$$\begin{aligned} \theta &= \text{Arccos}\left(\frac{\langle \vec{A}, \vec{B}\rangle}{\|\vec{A}\| \cdot \|\vec{B}\|}\right) = \text{Arccos}\left(\frac{0.999\,913\,209\,376}{1 \cdot 1}\right) \\ &= 1.317\,511\,893\,41 \times 10^{-2} \text{ radian.} \end{aligned}$$

(This angle corresponds to a distance of $6378 \times 1.317\,511\,893\,41 \times 10^{-2} \approx 84.031$ kilometers, or 52.214 miles.) Thus, the vector tangent at Abercombie Mountain to the great circle through both mountains is

$$\vec{g}'(1.317\,511\,893\,41 \times 10^{-2}) = \begin{pmatrix} 0.952\,336\,607\,019 \\ 0.153\,701\,481\,302 \\ 0.263\,497\,327\,452 \end{pmatrix}.$$

With $-\vec{g}'(1.317\,511\,893\,41 \times 10^{-2})$ pointing from Abercombie back toward Columbia, the azimuth becomes

$$\begin{aligned}\beta &= \pi + \operatorname{Arccos}\left(\frac{\vec{\mathbf{N}}_{\vec{\mathbf{B}}} \cdot \vec{g}'(1.317\,511\,893\,41 \times 10^{-2})}{\|\vec{\mathbf{N}}_{\vec{\mathbf{B}}}\| \cdot \|\vec{g}'(1.317\,511\,893\,41 \times 10^{-2})\|}\right) \\ &= \pi + \operatorname{Arccos}(0.401\,055\ldots) = \pi + 1.158\,128\ldots \text{ radian} \\ &= 180^\circ + 66^\circ 21' 21'' = 246^\circ 21' 21''.\end{aligned}$$

Observe that while the great circle $\vec{g}$ *through both mountains passes through Columbia with the azimuth* $65^\circ 34' 08''$, *it passes through Abercombie with the different azimuth* $66^\circ 21' 21''$.

5. Most great circles are *not* rhumblines. For example, consider the great circle $\vec{g}$ through the points $\vec{\mathbf{A}} = (1, 0, 0)$ and $\vec{\mathbf{B}} = (0, 1/\sqrt{2}, 1/\sqrt{2})$ on the unit sphere $S^2 \subset \mathbb{R}^3$. At $\vec{\mathbf{A}} = (1, 0, 0)$, the great circle $\vec{g}$ crosses the meridian at longitude $\varphi = 0$ at an angle equal to $\pi/4$, whereas at $\vec{\mathbf{B}} = (0, 1/\sqrt{2}, 1/\sqrt{2})$, the great circle $\vec{g}$ crosses the meridian at longitude $\varphi = \pi/2$ at an angle equal to $\pi/2$.

6. Proceed verbatim as in the solution to **Exercise 25**, with $b\vec{f}'(t)$ replaced by $\vec{w}$.

References

Ahlfors, Lars V. 1979. *Complex Analysis.* 3rd ed. New York: McGraw-Hill.

Audin, Michel, 1994. Courbes algébriques et systèmes intégrables: géodésiques des quadriques. *Expositiones Mathematicae* 1: 193–226.

Ayoub, Ayoub B. 1991. Proof without words: The reflection property of the parabola. *Mathematics Magazine* 64 (3) (June 1991): 175.

Bassein, Richard S. 1992. The length of the day. *American Mathematical Monthly* 99 (10) (December 1992): 917–921.

Brown, Lloyd A. 1979. *The Story of Maps.* New York: Dover.

Carey, Susan. 1996. Satellite navigation will require resurveying the world's airports. *Wall Street Journal* 134 No. 89 (6 May 1996) (Western Ed.) A9B.

Curjel, Caspar R. 1990. *Exercises in Multivariable and Vector Calculus.* New York: McGraw-Hill.

Edelman, Alan, and Eric Kostlan. 1995. How many zeroes of a random polynomial are real? *Bulletin of the American Mathematical Society* (N.S.) 32 (1) (January 1995): 1–37.

Fleming, Wendell. 1982. *Functions of Several Variables.* 2nd ed., corrected 2nd printing. New York: Springer-Verlag.

Goering, David K. 1990. Three families of map projections. Master's thesis. Cheney, WA: Eastern Washington University.

Lin, C. C., and L.A. Segel. 1988. *Mathematics Applied to Deterministic Problems in the Natural Sciences.* Philadelphia, PA: SIAM.

Liouville, Joseph. 1847. Note au sujet de l'article précédent [namely, Thomson [1847]]. *Journal de Mathématiques Pures et Appliquées* Série 1, 12 (juillet 1847): 265–290.

Maxwell, James Clark. 1892. *A Treatise on Electricity and Magnetism.* 3rd ed. Oxford, UK: Clarendon.

Richardus, Peter, and Ron K. Adler. 1972. *Map Projections: For Geodesists, Cartographers, and Geographers.* Amsterdam: North-Holland.

Reid, Miles. 1990. *Undergraduate Algebraic Geometry.* Cambridge, UK: Cambridge University Press.

Sachs, J.M. 1987. A curious mixture of maps, dates, and names. *Mathematics Magazine* 60: 151–158.

Shafarevich, I.R. 1977. *Basic Algebraic Geometry.* New York: Springer-Verlag.

Snyder, John P. 1987a. Bulletin 1532. Washington, DC: U.S. Government Printing Office.

Snyder, John P. 1987b. *Map Projections—A Working Manual.* Geological Survey Professional Paper 1395. Washington, DC: U.S. Government Printing Office.

Sommerfeld, Arnold. 1949. *Partial Differential Equations in Physics.* New York: Academic Press.

Spiegel, Murray S. 1964. *Complex Variables.* Schaum's Outline Series. New York: McGraw-Hill.

Spivak, Michael. 1980. *Calculus.* 2nd ed. Wilmington, DE: Publish or Perish.

Strang, Gilbert. 1991. *Calculus.* Wellesley, MA: Wellesley-Cambridge Press.

Thomson, William. 1847. Extraits de deux lettres adressées à M. Liouville. *Journal de Mathématiques Pures et Appliquées* Série 1, 12 (juillet 1847): 256–264.

Thurston, William P. 1994. On proof and progress in mathematics. *Bulletin of the American Mathematical Society* (N.S.) 30 (2) (April 1994): 161–177.

Tuchinsky, Phillip M. 1978. *Mercator's World Map and the Calculus.* UMAP Modules in Undergraduate Mathematics and Its Applications: Module 206. Lexington, MA: COMAP. Reprinted in *UMAP Modules: Tools for Teaching 1977–1979*, 677–727. Boston, MA: Birkhäuser, 1980.

U.S. Geological Survey. Undated. *Map Projections.* Anonymous poster. Reston, VA: U.S. Geological Survey (507 National Center, Reston, VA 22092, (800)–USA–MAPS).

Vest, Floyd, and Raymond Benge. 1994–1995. Latitude and longitude. *Consortium* HiMap Pull-Out Section. 52 (Winter 1994–1995).

van der Waerden, Bartel Leenert. 1983. *Geometry and Algebra in Ancient Civilizations.* New York: Springer-Verlag.

Weinstock, Robert. 1994. Isaac Newton: Credit where credit won't do. *Mathematics Magazine* 25 (3) (May 1994): 179–192.

Wolfram, Stephen. 1991. *Mathematica: A System for Doing Mathematics by Computer.* 2nd ed. Redwood City, CA: Addison-Wesley.

Yamada, Ken. 1993. Technology. *Wall Street Journal* (Western Ed.) 128 (81) (27 April 1993): B1.

Acknowledgments

The author gratefully acknowledges the following contributions. National Science Foundation grant DUE–9255539 supported the preparation of this Module and its first uses in workshops for mathematics instructors. Also, Dr. Dale A. Lear, formerly at Applied Geometry and now with McNeel & Associates in Seattle, presented to the workshop participants applications of differentials to the design of geometric software to compute intersections of surfaces for engineering and manufacturing purposes. Moreover, Boeing Computer Services' Geometry and Optimization Group, with Dr. Stephen P. Keeler (manager), Drs. David R. Ferguson, Thomas A. Grandine, Alan K. Jones, and Mr. Richard A. Mastro, have for several years demonstrated and discussed many applications of mathematics to the design of aircraft with students on campuses and with instructors in various workshops and conferences in the State of Washington. Furthermore, Prof. Gail Nord at Gonzaga University proofread several drafts of the manuscript and corrected many errors.

About the Author

Yves Nievergelt graduated in mathematics from the École Polytechnique Fédérale de Lausanne (Switzerland) in 1976, with concentrations in functional and numerical analysis of PDEs. He obtained a Ph.D. from the University of Washington in 1984, with a dissertation in several complex variables under the guidance of James R. King. He now teaches complex and numerical analysis at Eastern Washington University.

Prof. Nievergelt is an associate editor of *The UMAP Journal*. He is the author of several UMAP Modules, a bibliography of case studies of applications of lower-division mathematics (*The UMAP Journal* 6 (2) (1985): 37–56), and *Mathematics in Business Administration* (Irwin, 1989).

UMAP

Modules in Undergraduate Mathematics and Its Applications

Published in cooperation with the Society for Industrial and Applied Mathematics, the Mathematical Association of America, the National Council of Teachers of Mathematics, the American Mathematical Association of Two-Year Colleges, The Institute of Management Sciences, and the American Statistical Association.

Module 749

Fuzzy Sets

Elizabeth Youngstrom
Margaret Owens

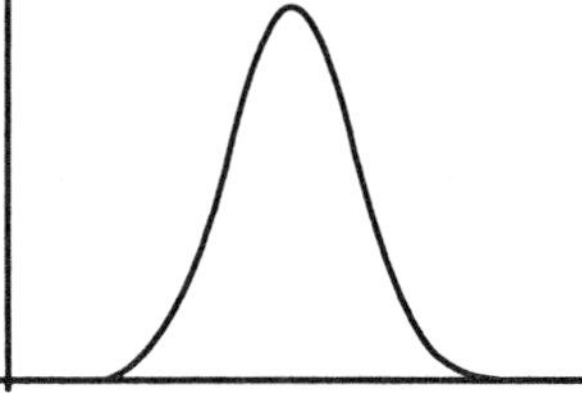

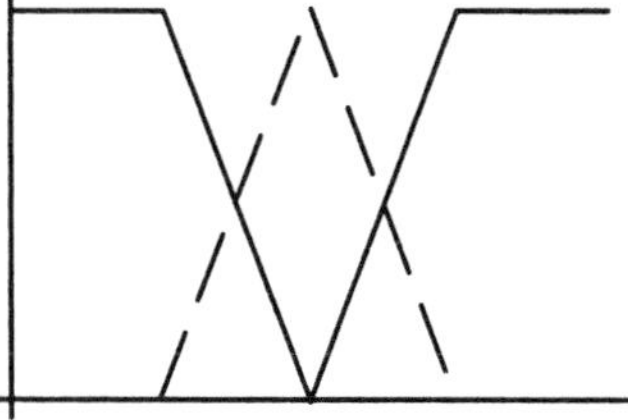

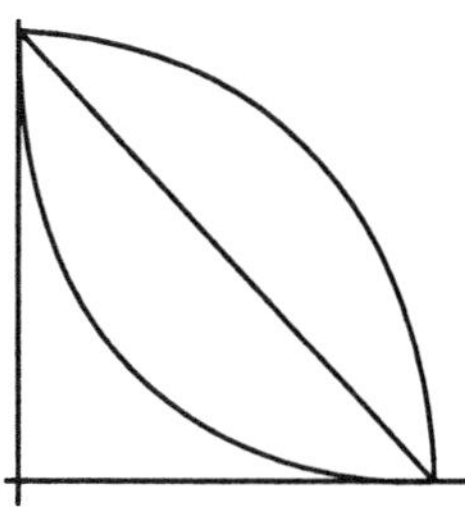

Applications of Discrete Mathematics and Set Theory to Computer Science and Engineering

COMAP, Inc., Suite 210, 57 Bedford Street, Lexington, MA 02173 (617) 862–7878

INTERMODULAR DESCRIPTION SHEET: UMAP Unit 749

TITLE: Fuzzy Sets

AUTHOR: Elizabeth Youngstrom
Department of Mathematics
Butte Community College
3536 Butte Campus Drive
Oroville, CA 95965

Margaret Owens
Department of Mathematics and Statistics
California State University, Chico
Chico, CA 95929–0525
mowens@oavax.csuchico.edu

MATHEMATICAL FIELD: Discrete mathematics, set theory

APPLICATION FIELD: Computer science, engineering

TARGET AUDIENCE: Students in a course in discrete mathematics

ABSTRACT: Fuzzy set theory is a relatively new area of mathematics that has been used very successfully in a variety of applications. In traditional set theory, an element is either in a set or not in it. In fuzzy set theory, every element of the universe of discourse is in every set to some extent. A fuzzy set is a collection of elements paired with numbers representing the element's level of membership in the fuzzy set. This Module explores elementary operations of set theory in a fuzzy setting and examines some applications of fuzzy set theory in decision-making.

PREREQUISITES:

- Familiarity with elementary set theory.
- Familiarity with function notation, including functions of two variables.
- An intuitive understanding of continuity (and also of differentiability, in one part of one problem).
- Familiarity with matrix multiplication would be helpful.

RELATED UNITS: Unit 752: *Fuzzy Relations*, by Elizabeth Youngstrom.

COMAP, Inc., Suite 210, 57 Bedford Street, Lexington, MA 02173
(800) 77–COMAP = (800) 772–6627 (617) 862–7878

Fuzzy Sets

Elizabeth Youngstrom
Dept. of Mathematics
Butte Community College
3536 Butte Campus Drive
Oroville, CA 95965

Margaret Owens
Dept. of Mathematics and Statistics
California State University, Chico
Chico, CA 95929–0525
mowens@oavax.csuchico.edu

Table of Contents

Modules and Monographs in Undergraduate Mathematics and its Applications (UMAP) Project

The goal of UMAP is to develop, through a community of users and developers, a system of instructional modules in undergraduate mathematics and its applications, to be used to supplement existing courses and from which complete courses may eventually be built.

The Project was guided by a National Advisory Board of mathematicians, scientists, and educators. UMAP was funded by a grant from the National Science Foundation and now is supported by the Consortium for Mathematics and Its Applications (COMAP), Inc., a nonprofit corporation engaged in research and development in mathematics education.

Paul J. Campbell	Editor
Solomon Garfunkel	Executive Director, COMAP

1. Introduction

Traditional mathematics is based on the ideas and principles of classical set theory, and classical set theory requires exactness. When we create a set in traditional mathematics, we look at each element from the universe and ask the yes or no question: "Does this element satisfy the characteristic properties of this set?" The element is put in the set if and only if the answer is yes. Unfortunately, in reality, many possible elements only *somewhat* satisfy the characteristic properties of the set, and it is difficult to know whether to answer yes or no. Many real-world concepts and questions have built-in ambiguity: Is Highway 99 "safe"? Does your friend have a "bad" cough? Is your cat "overweight"?

Let's examine the last question more carefully. Is your cat overweight? Suppose that your veterinarian tells you that a cat that weighs more than 15 pounds is overweight. Does this definition remove the ambiguity? Your cat weighs 14.75 lb. Is it safe to assume that your cat's weight is acceptable? Is there a black-and-white division between acceptable cat weights and cat weights that are too large? A standard mathematical model assumes the existence of such a boundary. Your cat weighs 14.75 lb and 14.75 is less than 15. If you believe in the standard model, don't worry about your cat's weight. Your cat is not overweight.

Mathematical models based on yes-or-no answers to such questions do not mirror the "gray" reality of real-world problems; the mathematical models must disguise the "gray" areas of the problem as either "black" or "white." When a real-world problem involves ambiguous circumstances, the best solution to the problem is arrived at when the uncertainties can be dealt with at each step of the problem-solving process. Fuzzy mathematics allows us to do just this by using "gray" mathematical models. In our fat cat example, your veterinarian might tell you that a 13-lb cat is slightly overweight, a 15-lb cat is overweight, and a 17-lb cat is grossly overweight. Moreover, every cat, including your 14.75-lb cat, lands in each of these three categories *to some degree.*

When we create a set in fuzzy set theory, our underlying assumption is that all elements of the universe are in the set to some extent. To create the fuzzy set, we look at each element of the universe and ask: "To what extent does this element satisfy the characteristic properties of this set?" Answers to this question can range from "completely" to "not at all." For each element of the universe, the answer to this question is converted into a number called a membership grade. A *fuzzy set,* therefore, contains all the elements of the universe, each paired with a number indicating the extent to which that element satisfies the characteristic properties of the set.

Fuzzy set theory allows us to describe collections of objects that do not have sharp boundaries. Some of the elementary ideas of fuzzy set theory can be traced back to theories of multivalued logic proposed in the early 1920s; however, it was another forty years before L. Zadeh [1965] laid the framework for fuzzy set theory. Since the publication of Zadeh's paper, fuzzy set theory has advanced from the status of a mathematical curiosity to a useful tool in mathematical modeling. Do you own an automobile with an anti-lock braking system? an autofocus camera? a color television? If so, you might own a consumer product that employs fuzzy technology.

Membership grades are a key ingredient of fuzzy mathematics; they are discussed in **Section 2**. When we change the structure of something as basic as "set," we must also examine the impact of the change on the other terms and operations of set theory. In fact, the familiar operations of set theory (complement, union, intersection) must be redefined. In **Section 3**, we examine set theoretic operations in a fuzzy setting; the result is *fuzzy set theory*. Then, in **Section 4**, we explore some of the surprising similarities and differences between classical set theory and fuzzy set theory. Set theory provides the foundation for most of classical mathematics. By stripping away the classical foundation and building on a fuzzy foundation, virtually any area of classical mathematics can be turned into fuzzy mathematics. In **Section 5**, we see how the notion of a relation between sets can be "fuzzified."

In the following sections, when we refer to "fuzzy" mathematics, we are by no means implying that the mathematics is imprecise. Rather, the "fuzz" arises from making the ambiguity or "fuzziness" found in the real world a part of the mathematical model. The mathematical tools and concepts we apply to fuzzy models are just as precise as the traditional mathematical tools and concepts we apply to standard mathematical models.

2. Membership in Fuzzy Sets

2.1 Membership Grades

A basic underlying assumption of fuzzy set theory is that all elements of the universal set are elements of all fuzzy sets in that universe. To define a fuzzy set **F** in a fixed traditional universal set U, every member of U is associated with a *membership grade*, a number between 0 and 1 representing the degree to which the element belongs in **F**. If the element has none of the characteristics of the collection **F**, then it has a membership grade of 0 in **F**. If there is no doubt in your mind that the element does have the characteristics of **F**, then it receives a membership grade of 1 in **F**. However, as mentioned earlier, most real-world situations are not that clear. As a result, most elements of U belong to **F** to some degree between 0 and 1. Elements with membership grades closer to 1 are the elements that an expert feels most confident about placing in **F**. Those closer to 0 are the ones that an expert is not very comfortable associating with **F**'s characteristics. Membership grades allow the mathematical model to reflect the ambiguity of the real world. (Throughout the Module, we use upper-case italic letters to represent traditional sets and upper-case bold letters to represent fuzzy sets.)

Let's play "expert." Suppose that your company reassigns you to work at their office in TinyTown, USA. When you get there, you quickly discover there are only 7 homes for sale in TinyTown. We label the homes a, b, c, d, e, f, and g. The set of available homes $\{a, b, c, d, e, f, g\}$ is the universal set for this example. Let's talk about the collection of "beautiful" homes in TinyTown. The fuzzy set of beautiful homes is a set consisting of the elements of the universe $\{a, b, c, d, e, f, g\}$ paired with their corresponding membership grades, numbers between 0 and 1 representing the degree to which each home is "beautiful." Since the term "beautiful" is quite ambiguous, the membership grades are likely to be assigned somewhat differently by different "experts." If you were working on a complex fuzzy engineering design, you would want the most experienced and knowledgeable individuals in the field helping you to set up your membership assignments for the elements of your fuzzy sets. In our example, since you are the expert, you set up the membership assignments. As you look at each home, ask yourself: "To what extent is this home beautiful?" The symbol used to pair an element with its membership grade is a vertical slash, |. Suppose that you decide that home a has a membership grade of .8 in the fuzzy set of beautiful homes, **B**; then the pair $a|.8$ appears in the list of members of **B**. Each home in the universal set must be assigned such a membership grade. The result might be the following fuzzy set:

$$\mathbf{B} = \text{fuzzy set of beautiful homes} = \{a|.8,\ b|.3,\ c|1,\ d|0,\ e|.5,\ f|0,\ g|.1\}.$$

That is, you thought that home a was really quite beautiful; so you gave it a grade of .8. Home b had a small beautiful stained-glass window, but the rest of the house was not very beautiful. Consequently, home b got a membership grade of .3. Home c was the most beautiful home you could imagine; so, of course, it received a membership grade of 1. Home d, on the other hand, was just plain ugly. You saw no beauty in it whatsoever. Thus, home d received a membership grade of 0 in the set of beautiful homes. Homes e, f, and g were assigned their membership grades similarly.

On the other hand, if we limit ourselves to classical set theory and look at the classical or *crisp set* of beautiful homes, then as you look at each home, you ask yourself: "Is this home beautiful?" Your only choices for answers are "yes" and "no." If you answer "yes" for homes a, c and e and "no" for the other homes, then the set of beautiful homes is $\{a, c, e\}$. Notice there is no distinction here between the home that you felt was the most beautiful home in the world and the home that was only somewhat beautiful. Herein lies the greatest advantage of using a fuzzy-set-theoretic foundation as opposed to traditional set theory. The distinction between home c and home e is a part of the fuzzy mathematical model, but that distinction is lost in a traditional mathematical model. The traditional model contains no clue about the distinction; information has been lost in the process of creating the model. However, subsequent operations and strategies used in the fuzzy model automatically take the difference between homes c and e into account; information about the difference between the two homes is retained and used in the fuzzy model.

The temptation to equate membership grades with probabilities is strong, presumably because in both fuzzy set theory and probability theory each element of a universal set is associated with a number between 0 and 1. However, the differences between fuzzy set theory and probability theory are quite striking, whether you look at the underlying concepts for each or the outward appearance of their mathematical models. Perhaps the most obvious difference between probabilities and membership grades is that the sum of the probabilities for all possible outcomes must add up to 1, whereas the membership grades in a fuzzy set very seldom add up to 1. But the most significant difference lies in the very natures of the two theories. Probability is based on the concept of randomness: Sometimes an event occurs; sometimes it does not. The probability of an event predicts the *likelihood* that the event will occur. Membership grades in fuzzy set theory refer to the *extent* to which an event has occurred. In our TinyTown example, the .8 membership grade for home a is not the probability of picking home a based on beauty. It is the extent to which home a is beautiful.

Exercises

1. a) When you graduate from college, suppose you get four job offers: company a offers \$45,000; company b offers \$35,000; company c offers \$30,000; company d offers \$15,000. Form a crisp set C of companies offering well-paying jobs.

b) Using the data given in **a**, form a fuzzy set **F** of companies offering well-paying jobs.

2. a) Based solely on the salaries given in **Exercise 1a**, determine possible values for the probabilities of choosing each of the jobs.

b) How do these probabilities compare with the membership grades for each element in your fuzzy set **F**?

2.2 The Membership Function

At this point, we need to take a closer look at the mathematics of assigning membership grades. We begin with a crisp universal set U. In a given fuzzy set S, each element of U is associated with exactly one number from the unit interval $[0, 1]$. The assignment of the number to the element is governed by the *membership function*. This function maps elements from the universe U into the interval [0,1]. If we let μ_{S} denote the membership function for fuzzy set S, we have $\mu_S : U \to [0, 1]$. For each element x in set U, let $\mu_{\mathrm{S}}(x)$ represent the membership grade of x in the fuzzy set **S**. This membership grade results from the answer to the question: To what extent does x have the characteristics specified by set S?

Our TinyTown example has a finite universal set, but a fuzzy set can have an infinite universal set. For example, we could let the universal set be the set of real numbers and look at **H**, the fuzzy set of *numbers close to 100*. A graphical representation of the membership function μ_{H} should show that the closer a number is to 100, the higher is its membership grade in **H**. There are infinitely many reasonable choices for such a function. Piecewise-linear functions, like that in **Figure 1**, are often used in applications of fuzzy set theory. These functions are easy to draw, easy to compute with, and usually provide as good a model of the physical situation as a more complicated nonlinear function like that in **Figure 2**.

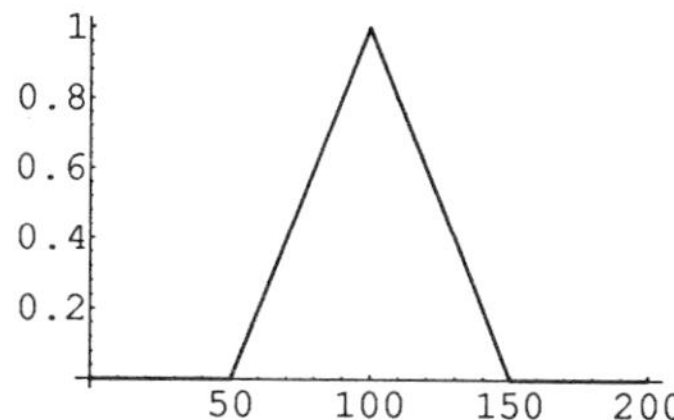

Figure 1. The membership function for **H**, the fuzzy set of numbers close to 100.

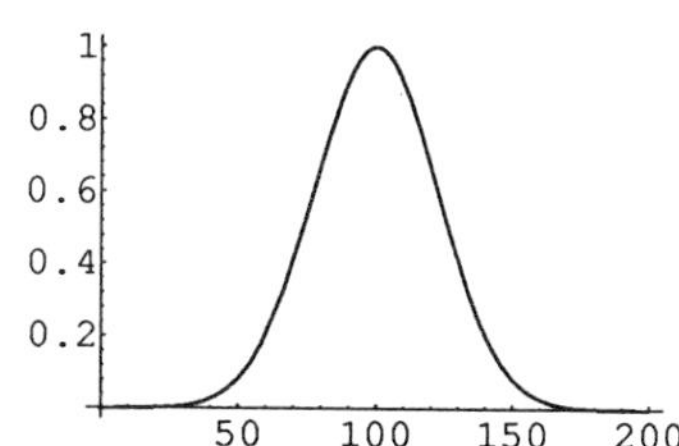

Figure 2. Another possible membership function for fuzzy set **H**.

In **Section 3**, we will use the membership function in **Figure 1** for our fuzzy set **H**. This function is defined as follows:

$$\mu_{\mathbf{H}}(x) = \begin{cases} 0, & \text{if } x \in (-\infty, 50); \\ 0.02x - 1, & \text{if } x \in [50, 100]; \\ -0.02x + 3, & \text{if } x \in (100, 150); \\ 0, & \text{if } x \in [150, +\infty). \end{cases}$$

What are the consequences of "fuzzifying" sets? What happens to the rest of set theory? For example, when are two fuzzy sets **R** and **S** equal? In traditional set theory, two sets R and S are equal if and only if they have exactly the same elements. But in fuzzy set theory, every fuzzy set contains every element of the universal set to some extent. What defines the difference between two fuzzy sets **R** and **S**? The difference, if any, lies not in the lists of elements in the fuzzy sets, but in the fuzzy sets' membership functions. Fuzzy set **R** is *equal* to fuzzy set **S** if and only if each element's membership grade in **R** is the same as its membership grade in **S**:

$$\mathbf{R} = \mathbf{S} \text{ if and only if } \mu_{\mathbf{R}}(x) = \mu_{\mathbf{S}}(x) \text{ for all } x \in U.$$

We need membership functions to define the notion of equality of fuzzy sets. In **Section 3**, we will see further evidence that membership functions are a key ingredient in fuzzy set theory.

Exercises

3. Sketch the graph of a possible membership function for each of the following fuzzy sets of real numbers.

a) the fuzzy set of numbers that are close to 4

b) the fuzzy set of numbers that are much larger than 4

c) the fuzzy set of numbers whose square is not close to 4

4. Sketch the graph of a possible membership function for the fuzzy set of acceptable annual starting salaries for a recent college graduate. Let the universal set be the interval [\$0, \$100000] and use your experience to select membership levels for elements of the universal set.

3. Operations on Fuzzy Sets

Fuzzy set theory has the same three major operations as classical set theory: complement, union, and intersection. That same concept of set membership that demanded a change in the definition of equal sets also demands that we look differently at these three operations. In fuzzy set theory, all elements of the universe are in every fuzzy set in the universe to some degree. Therefore, all elements of a set are also in every set's complement, as well as its union and its intersection with other fuzzy sets. In fuzzy set theory, we never lose or gain elements; only the membership levels change. Each fuzzy operation is defined by a function that operates on the membership functions of the fuzzy sets involved.

3.1 Complement

The *complement* $\bar{\mathbf{R}}$ of a fuzzy set $\mathbf{R}$ is defined by a function f that acts on the membership grade of each element of $\mathbf{R}$. That is, $\mu_{\overline{R}}(x) = f\big(\mu_R(x)\big)$ for some function $f : [0,1] \rightarrow [0,1]$. To avoid possible confusion of the complement function with the membership functions, henceforth we refer to f as an *operator*. The operator f is a function that acts on membership functions and defines the complement operation. Unlike classical set theory with its unique definition of complement, there are many possible choices for the complement function in fuzzy set theory.

Going back to our TinyTown example, the complement of our fuzzy set of beautiful homes $\mathbf{B}$ is the fuzzy set of **not** *beautiful homes* $\overline{\mathbf{B}}$. Since home a was given a membership grade of .8 in $\mathbf{B}$, it must be quite beautiful. This implies that home a should have a low membership grade in the set of homes that are not beautiful (maybe .2?). On the other hand, home d with its membership grade of 0 in $\mathbf{B}$ was just plain ugly. Therefore, its membership grade in the set of homes that are not beautiful should be very high (maybe even 1?). We need an operator that gives reasonable membership grades in the complement. If an element has a high membership grade in the original set, then it should have a low membership grade in the complement of that set, and vice versa. One way to be sure this happens is to subtract the element's membership grade from 1 to get its membership grade in the complement; that is, let

$$\mu_{\overline{\mathbf{B}}}(x) = 1 - \mu_{\mathbf{B}}(x) \text{ for all } x \in U.$$

The operator in question, $f(r) = 1 - r$ for all $r \in [0,1]$, is called the *standard complement operator*. Using this complement operator, fuzzy set $\overline{\mathbf{B}}$ is created, element by element:

$$
\begin{aligned}
\mu_{\overline{\mathbf{B}}}(a) &= 1-\mu_{\mathbf{B}}(a) &= 1-.8 &= .2,\\
\mu_{\overline{\mathbf{B}}}(b) &= 1-\mu_{\mathbf{B}}(b) &= 1-.3 &= .7,\\
\mu_{\overline{\mathbf{B}}}(c) &= 1-\mu_{\mathbf{B}}(c) &= 1-1 &= 0,
\end{aligned}
$$

and so forth. The complement is $\overline{\mathbf{B}} = \{a|.2,\ b|.7,\ c|0,\ d|1,\ e|.5,\ f|1,\ g|.9\}$.

Although the standard complement operator is used most frequently, many other entire classes of complement operators have been proposed and found to be useful in different real world situations. (See **Exercise 6.**) Having an entire class of operators to choose from when setting up your fuzzy mathematical model allows you to be as conservative as you want with your data when defining the complement. To ensure intuitively pleasing results, these complement operators have a few conditions placed on them. To be called a complement operator, a function $f : [0,1] \to [0,1]$ must satisfy at least the first two of the following four properties.

1. Function f must satisfy the *boundary condition*. This condition defines activity at the two ends of the interval $[0,1]$ by requiring $f(0) = 1$ and $f(1) = 0$.

2. Function f must be *monotone nonincreasing*. That is, for all $r, r' \in [0,1]$, if $r \le r'$, then $f(r) \ge f(r')$. In terms of membership grades, if f is monotone nonincreasing and $\mu_{\mathbf{R}}(x) \le \mu_{\mathbf{R}}(y)$, then $\mu_{\overline{\mathbf{R}}}(x) \ge \mu_{\overline{\mathbf{R}}}(y)$ for all $x, y \in U$. Intuitively, if element x satisfies the characteristic properties of the collection $\mathbf{R}$ less than element y, then element x satisfies the characteristic properties of the complement of $\mathbf{R}$ more than y.

If you look at the membership grades of any two homes in our fuzzy set $\mathbf{B}$, the one with the larger membership grade, the more beautiful home, has the smaller membership grade in $\overline{\mathbf{B}}$. For example, since home a is more beautiful than home b, it follows that home a is less "not beautiful" than home b.

Two further restrictions on f are almost always included.

3. Function f must be *continuous*. Intuitively, this guarantees that a small difference in membership grades in a particular fuzzy set does not result in a large difference in membership grades in the complement of the fuzzy set. In our example, home f and home g have close to the same membership grades in $\mathbf{B}$, 1 and .9 respectively; so their membership grades in $\overline{\mathbf{B}}$, 0 and .1 respectively, are also close in value.

4. Function f must be *involutive*. That is, for all $r \in [0,1]$, we must have $f(f(r)) = r$. Equivalently, f must be its own inverse under composition. In terms of fuzzy sets, for any fuzzy set $\mathbf{R}$, we have $\overline{\overline{\mathbf{R}}} = \mathbf{R}$: The complement of the complement of a fuzzy set is the fuzzy set. In terms of membership grades, for all $x \in U$, we have $\mu_{\overline{\overline{\mathbf{R}}}}(x) = \mu_{\mathbf{R}}(x)$: The membership grade of x in the complement of the complement of $\mathbf{R}$ is the same as the membership grade of x in $\mathbf{R}$.

Some examples of nonstandard complement operators that satisfy all four of these conditions are described in **Exercise 6** below.

Recall our infinite fuzzy set **H** of numbers close to 100. The dotted graph in **Figure 3** is our membership function for **H**. The solid graph in **Figure 3** is the membership function for $\overline{\mathbf{H}}$, obtained by using the standard complement operator.

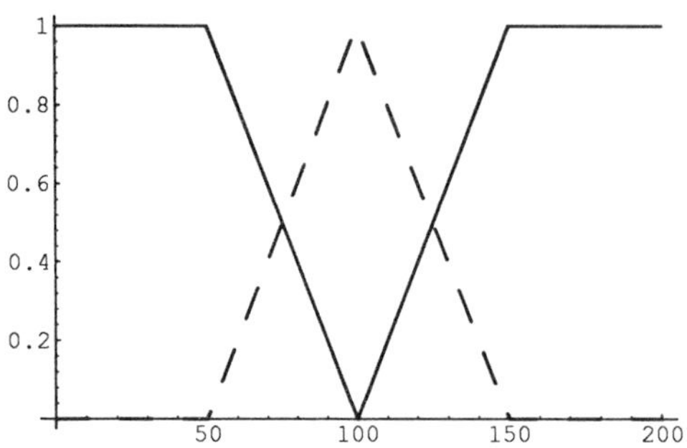

Figure 3. Membership functions for fuzzy set **H** and its complement.

Algebraically, the membership function for the complement of **H** is the following function:

$$\mu_{\overline{H}}(x) = 1 - \mu_H(x) = \begin{cases} 1, & \text{if } x \in (-\infty, 50); \\ -0.02x + 2, & \text{if } x \in [50, 100]; \\ 0.02x - 2, & \text{if } x \in (100, 150); \\ 1, & \text{if } x \in [150, +\infty). \end{cases}$$

Exercises

5. Find the complement of the following fuzzy sets using the standard complement operator:
   a) $\mathbf{Y} = \{\otimes|.36,\ \triangle|.09,\ \square|0,\ \diamondsuit|1,\ \nabla|.64\}$
   b) $\mathbf{Z} = \{x|\frac{1}{x} : x \geq 1\}$. In this set, each real number greater than or equal to 1 has a membership grade of $1/x$ in **Z**. For example, 2 has membership grade 1/2.

6. The *Yager complement operators* are a class of nonstandard complement operators. They define an infinite collection of complement operations by defining a complement operator, f_w, for each $w \in (0, \infty)$. In particular, for each $w \in (0, \infty)$ and each $r \in [0, 1]$, they let $f_w(r) = (1 - r^w)^{1/w}$. It follows that the membership grade of an element x in the complement of a fuzzy set **R** is given by $\mu_{\overline{R}}(x) = \left[1 - \left(\mu_R(x)\right)^w\right]^{1/w}$. Note that when $w = 1$, the resulting function is the standard complement operator. See **Figure 4** for sketches of $f_{.5}$, f_1, and f_2.

a) Find the Yager complements of sets **Y** and **Z** from **Exercise 5** using $w = .5$.

b) Repeat part **a** using $w = 2$. For $\overline{\mathbf{Y}}$, express membership grades to the nearest hundredth.

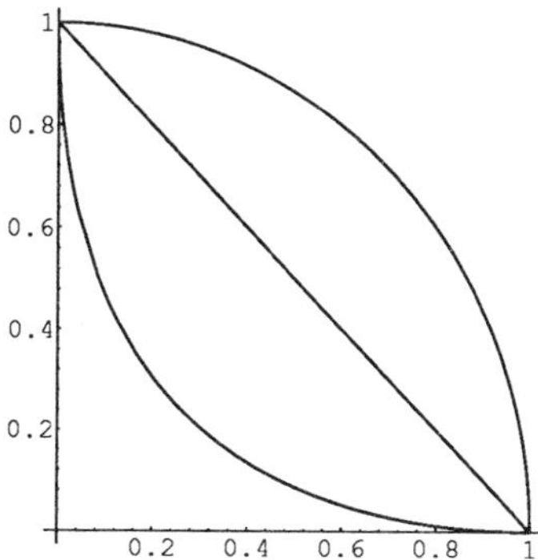

Figure 4. The Yager complement operators $f_{.5}$, f_1, and f_2.

7. Show that the Yager complement operators satisfy the four properties for complement operators mentioned earlier.

8. Sketch the graph of a non-Yager complement operator that satisfies all four complement properties. (Hint: all Yager complement operators are differentiable functions.)

3.2 Union

The *union* of two fuzzy sets **R** and **S** is another fuzzy set, denoted **R** ∪ **S**, which is defined by a union operator g that assigns a new membership grade to each element of the universe by comparing the element's membership grades in the two original fuzzy sets. That is,

$$\mu_{\mathbf{R}\cup\mathbf{S}}(x) = g(\mu_{\mathbf{R}}(x), \mu_{\mathbf{S}}(x))$$

for some function $g : [0,1] \times [0,1] \to [0,1]$. The set $[0,1] \times [0,1]$ is called the *Cartesian product* of $[0,1]$ and $[0,1]$. It is the set of all ordered pairs whose components both lie in $[0,1]$; that is,

$$[0,1] \times [0,1] = \{(r,s) : r \in [0,1] \text{ and } s \in [0,1]\}.$$

We will see that, as with the complement, there are many choices for the union operator. The following discussion highlights the natural choice for union operator.

Returning to our TinyTown example, recall that

$$\mathbf{B} = \text{fuzzy set of beautiful homes } = \{a|.8,\ b|.3,\ c|1,\ d|0,\ e|.5,\ f|0,\ g|.1\}.$$

Suppose that we construct another fuzzy set of available homes: the fuzzy set of affordable homes defined by

$$\mathbf{A} = \text{fuzzy set of affordable homes } = \{a|.4,\ b|.7,\ c|0,\ d|.9,\ e|.4,\ f|.8,\ g|1\}.$$

The fuzzy set $\mathbf{B} \cup \mathbf{A}$ corresponds logically to the set of *beautiful* **or** *affordable homes*. The function that assigns the membership grade to home a in this union of fuzzy sets must look at the membership grades for home a in the fuzzy set of beautiful homes and in the fuzzy set of affordable homes, .8 and .4 respectively. Since we are looking at beautiful or affordable homes, home a should have a membership grade at least as large as its largest membership grade in those two fuzzy sets. If we wish to be conservative, we would keep the membership grade in the union as low as possible, subject to this condition. The membership grade in the union is then the maximum of the membership grades in the two original fuzzy sets:

$$\mu_{\mathbf{B}\cup\mathbf{A}}(x) = \max\{\mu_{\mathbf{B}}(x), \mu_{\mathbf{B}}(x)\}.$$

Using this union operator, the membership grades for the elements of $\mathbf{B} \cup \mathbf{A}$ are determined, element by element:

$$\begin{aligned}
\mu_{\mathbf{B}\cup\mathbf{A}}(a) &= \max\{\mu_{\mathbf{B}}(a), \mu_{\mathbf{A}}(a)\} = \max\{.8, .4\} = .8, \\
\mu_{\mathbf{B}\cup\mathbf{A}}(b) &= \max\{\mu_{\mathbf{B}}(b), \mu_{\mathbf{A}}(b)\} = \max\{.3, .7\} = .7, \\
\mu_{\mathbf{B}\cup\mathbf{A}}(c) &= \max\{\mu_{\mathbf{B}}(c), \mu_{\mathbf{A}}(c)\} = \max\{1, 0\} = 1,
\end{aligned}$$

and so forth. The resulting union is

$$\begin{aligned}
\mathbf{B} \cup \mathbf{A} &= \text{fuzzy set of beautiful or affordable homes} \\
&= \{a|.8,\ b|.7,\ c|1,\ d|.9,\ e|.5,\ f|.8,\ g|1\}.
\end{aligned}$$

The union operator that we have just discovered is the *standard union operator*. Given fuzzy sets **R** and **S**, this operator, $g(r, s) = \max\{r, s\}$, assigns membership grades to the elements in $\mathbf{R} \cup \mathbf{S}$ as follows:

$$\mu_{\mathbf{R}\cup\mathbf{S}}(x) = \max\{\mu_{\mathbf{R}}(x), \mu_{\mathbf{S}}(x)\} \text{ for all } x \in U.$$

Although this standard union operator is used most often, many other whole classes of union operators have been developed (for example, see **Exercise 11**). The advantage of choice once again is the flexibility you are afforded when setting up your fuzzy mathematical model. How much do you want the lower membership grade to contribute to the resulting union? How strong or dominant should the larger membership grade be? To some extent, you may choose these levels of contribution. However, there are some strings attached.

The conditions below assure that the function $g : [0,1] \times [0,1] \to [0,1]$ behaves as we would expect a union operator to behave.

1. Function g must satisfy a *boundary condition*: $g(r, 0) = r$ for all $r \in [0,1]$. When element x has membership level 0 in **S**, the element's membership level in $\mathbf{R} \cup \mathbf{S}$ is equal to its membership level in **R**.

2. Function g must be *commutative;* that is, $g(r,s) = g(s,r)$ for all $r, s \in [0,1]$. Since it follows that $\mu_{\mathbf{R}\cup\mathbf{S}}(x) = \mu_{\mathbf{S}\cup\mathbf{R}}(x)$ for all $x \in U$, this condition ensures that $\mathbf{R} \cup \mathbf{S} = \mathbf{S} \cup \mathbf{R}$.

3. Function g must be *monotone nondecreasing;* that is, if $r \leq r'$ and $s \leq s'$, then $g(r,s) \leq g(r',s')$. In terms of membership grades, for all $x \in U$, if $\mu_{\mathbf{R}}(x) \leq \mu_{\mathbf{R}'}(x)$ and $\mu_{\mathbf{S}}(x) \leq \mu_{\mathbf{S}'}(x)$, then $\mu_{\mathbf{R}\cup\mathbf{S}}(x) \leq \mu_{\mathbf{R}'\cup\mathbf{S}'}(x)$. If the membership grade of x increases in either $\mathbf{R}$ or $\mathbf{S}$, then the membership grade of x in $\mathbf{R} \cup \mathbf{S}$ does not decrease.

4. Function g must be *associative;* that is, for all $r, s, t \in [0,1]$, we must have $g\big(r, g(s,t)\big) = g\big(g(r,s), t\big)$. This property implies associativity of the fuzzy union operation: $\mathbf{R} \cup (\mathbf{S} \cup \mathbf{T}) = (\mathbf{R} \cup \mathbf{S}) \cup \mathbf{T}$.

Any operator $g : [0,1] \times [0,1] \to [0,1]$ that satisfies properties 1 through 4 is called a *union operator.* The standard union operator has all four of these properties and two additional properties. No other union operator has both of these additional properties:

5. Function g is *continuous.* Intuitively, when the union operator is continuous, a small change in the membership grade of element x in either $\mathbf{R}$ or $\mathbf{S}$ cannot result in a large change in the membership grade of x in $\mathbf{R} \cup \mathbf{S}$.

6. Function g is *idempotent;* that is, $g(r,r) = r$ for all $r \in [0,1]$. This property ensures that $\mathbf{R} \cup \mathbf{R} = \mathbf{R}$ holds for all fuzzy sets $\mathbf{R}$.

The combination of these two additional characteristics is many times the factor making the standard union operator the best choice for a mathematical model. It is, perhaps, surprising that only one function satisfies all six of these properties. Therefore, we provide the following proof of this fact. We begin with a helpful lemma.

Lemma. *If function $g : [0,1] \times [0,1] \to [0,1]$ satisfies union properties 1, 2, and 3, then $g(r,s) \geq \max\{r,s\}$ for all $r, s \in [0,1]$.*

Proof: Assume $r, s \in [0,1]$. It suffices to verify that $g(r,s) \geq r$ and $g(r,s) \geq s$. Since $s \geq 0$, by the monotone condition, $g(r,s) \geq g(r,0)$. By the boundary condition, $g(r,0) = r$. Combining these two results gives $g(r,s) \geq r$. To obtain the second inequality, we first use the commutativity of g, and then use the monotone and the boundary conditions on g:

$$g(r,s) = g(s,r) \geq g(s,0) \geq s. \qquad \square$$

Next, we use the lemma to prove our claim about the standard union operator.

Theorem 1. *The standard union operator is the only function that satisfies all six union properties.*

Proof: Let $g : [0,1] \times [0,1] \to [0,1]$ be a function that satisfies all six union properties. Let r and s be in $[0,1]$. By the **Lemma,** $g(r,s) \geq \max\{r,s\}$; we must prove that $g(r,s) \leq \max\{r,s\}$ also holds. Suppose, without loss of generality, that $r \leq s$. Since g is monotone, $g(r,s) \leq g(s,s)$. Since g is idempotent, $g(s,s) = s$. Consequently, $g(r,s) \leq s = \max\{r,s\}$. □

Notice that in our proofs of the **Lemma** and **Theorem 1,** we did not use union properties 4 or 5. So, in fact, we have shown that there is only one function that satisfies properties 1, 2, 3, and 6. That function, $g(r,s) = \max\{r,s\}$, also happens to be both associative and continuous.

Our TinyTown example involves a finite universal set. The standard union operator behaves just as nicely in infinite universal sets. Let us consider a union of two infinite fuzzy sets. Fuzzy set **H**, the fuzzy set of numbers close to 100, was defined in **Section 2;** the membership function for **H** is the dotted graph in **Figure 5**. Suppose that the membership grades for another fuzzy set **K**, the fuzzy set of numbers close to 150, are given by the solid graph in **Figure 5**. Fuzzy set **K** is defined by the following membership function:

$$\mu_K(x) = \begin{cases} 0, & \text{if } x \in (-\infty, 100); \\ 0.02x - 2, & \text{if } x \in [100, 150]; \\ -0.02x + 4, & \text{if } x \in (150, 200); \\ 0, & \text{if } x \in [200, +\infty). \end{cases}$$

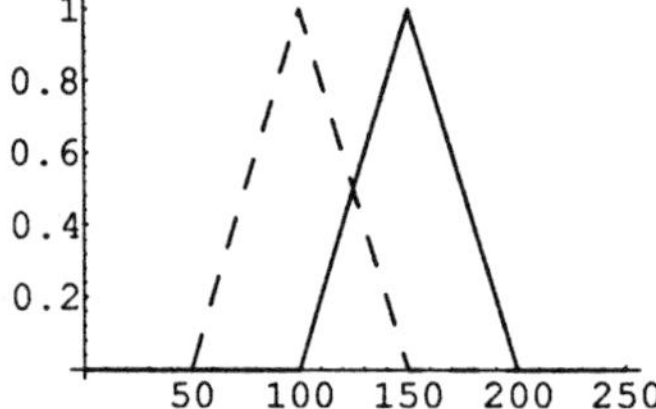

Figure 5. Membership functions for fuzzy sets **H** and **K**.

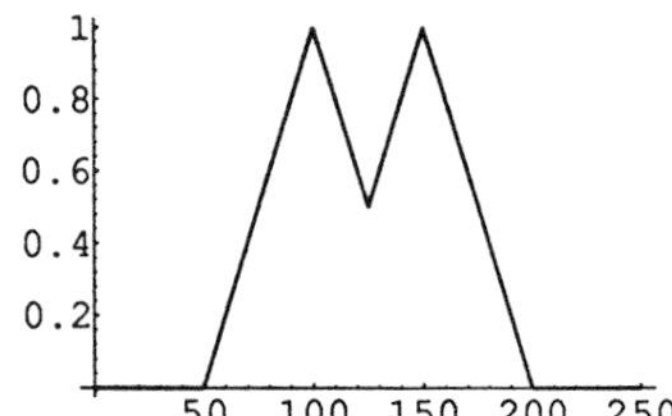

Figure 6. Membership function for **H**∪**K**, the fuzzy set of numbers close to either 100 or 150.

The solid graph in **Figure 6** is the membership function for the union of **H** and **K** , the fuzzy set of numbers close to either 100 or 150, using the standard union operator. The x-value at which the graphs of $\mu_{\mathbf{H}}$ and $\mu_{\mathbf{K}}$ intersect lies in the interval [100,150]. We can find the exact value at which these graphs cross by setting the expressions $(-0.02x + 3)$ and $(0.02x - 2)$ equal and solving for x; the result is $x = 125$. With this value, we can explicitly describe the membership function for $\mathbf{H} \cup \mathbf{K}$:

$$\mu_{\mathbf{H}\cup\mathbf{K}}(x) = \max\{\mu_{\mathbf{H}}(x), \mu_{\mathbf{K}}(x)\} = \begin{cases} 0, & \text{if } x \in (-\infty, 50); \\ \;\;0.02x - 1, & \text{if } x \in [50, 100]; \\ -0.02x + 3, & \text{if } x \in (100, 125); \\ \;\;0.02x - 2, & \text{if } x \in [125, 150]; \\ -0.02x + 4, & \text{if } x \in (150, 200); \\ 0, & \text{if } x \in [200, +\infty). \end{cases}$$

Exercises

9. Recall fuzzy sets **Y** and **Z** from **Exercise 5**:

$$\mathbf{Y} = \{\otimes|.36,\ \triangle|.09,\ \square|0,\ \Diamond|1,\ \nabla|.64\}, \qquad \mathbf{Z} = \{x|\tfrac{1}{x} : x \geq 1\}.$$

Let

$$\mathbf{T} = \{\otimes|.3,\ \triangle|0,\ \square|.8,\ \Diamond|.9,\ \nabla|.5\}, \qquad \mathbf{V} = \{x|\tfrac{x}{x+1} : x \geq 1\}.$$

a) Using the standard union operator, find $\mathbf{T} \cup \mathbf{Y}$.

b) Using the standard union operator, find $\mathbf{V} \cup \mathbf{Z}$, and sketch the graph of the membership function.

10. In this section, we proved that max is the only function that satisfies all six of the union properties. Nonetheless, some believe that max is not intuitively appealing as a union function. The problem is that for each element x of the universal set, x's membership level in $\mathbf{R} \cup \mathbf{S}$ is one of $\mu_{\mathbf{R}}(x)$ and $\mu_{\mathbf{S}}(x)$; after the maximum value is identified, the other value makes no further contribution to the membership grade of x in the union. The fact remains that any other function we try fails to satisfy at least one of the six union properties. For the functions $g : [0,1] \times [0,1] \to [0,1]$ defined below, determine which of the six union properties fail to hold.

a) $g(r,s) = \min\{r+s, 1\}$

b) $g(r,s) = (r+s)/2$

c) $g(r,s) = 1 - \sqrt{(1-r)(1-s)}$

d) $g(r,s) = r + s - rs$

11. One class of union operators, called the *Yager union operators*, defines an operator function $g_w(r,s) = \min\{1, (r^w + s^w)^{1/w}\}$, for each $w \in (0, \infty)$. As w takes on different values, we get different union operators. Let **R** and **S** be two fuzzy sets. Let $r = \mu_{\mathbf{R}}(x)$ and $s = \mu_{\mathbf{S}}(x)$ for some $x \in U$. Then, according to this definition, for a fixed $w \in (0, \infty)$,

$$\mu_{\mathbf{R}\cup\mathbf{S}}(x) = g_w(r,s) = \min\{1, (r^w + s^w)^{1/w}\}.$$

Note that the function defined in **Exercise 10a** is the Yager union operator g_1.

a) Using fuzzy sets **T** and **Y** defined in **Exercise 9**, find $\mathbf{T} \cup \mathbf{Y}$ using Yager's union operator with $w = 0.5$. (Express membership grades to the nearest hundredth.)

b) Repeat part **a** using $w = 2$.

12. a) Show that all Yager union operators satisfy the boundary condition.

b) Show that all Yager union operators are commutative.

c) Show that *no* Yager union operator is idempotent. (This is the only one of the six union properties that Yager operators fail to satisfy.)

3.3 Intersection

The *intersection* of two fuzzy sets **R** and **S** is another fuzzy set denoted **R** ∩ **S**. The intersection, like the union, is defined by an operator; the operator determines how membership grades are assigned in the intersection. That is, $\mu_{\mathbf{R}\cap\mathbf{S}}(x) = h(\mu_{\mathbf{R}}(x), \mu_{\mathbf{S}}(x))$ for some function $h : [0,1] \times [0,1] \to [0,1]$. Many functions have been used to define the work of the intersection operation but, again, we will find there is a standard choice for the operator.

In our TinyTown example, recall that we have two fuzzy sets:

B = fuzzy set of beautiful homes = $\{a|.8,\ b|.3,\ c|1,\ d|0,\ e|.5,\ f|0,\ g|.1\}$,

A = fuzzy set of affordable homes = $\{a|.4,\ b|.7,\ c|0,\ d|.9,\ e|.4,\ f|.8,\ g|1\}$.

Let's look at two more fuzzy sets for our house-hunting scenario, the fuzzy sets of homes that are in a good location and homes that are well constructed. Let

L = fuzzy set of *homes in a good location* = $\{a|.6,\ b|0,\ c|1,\ d|.2,\ e|0,\ f|0,\ g|.7\}$,

C = fuzzy set of *well-constructed homes* = $\{a|.5,\ b|.6,\ c|.9,\ d|.7,\ e|.6,\ f|0,\ g|.8\}$.

The logic operator *and* corresponds to intersection in set theory; therefore, the fuzzy set of homes that are *beautiful* **and** *in a good location* is the intersection of **B** and **L**, or **B** ∩ **L**. To assign a membership grade to home *a* in this fuzzy set, we must look at the membership grades for home *a* in the fuzzy set of beautiful homes and in the fuzzy set of homes in a good location, .8 and .6 respectively. Since we want home *a* to have both properties, the membership grade for home *a* in the intersection should not be larger than its smallest membership grade in either of the two original fuzzy sets. Suppose we let the membership grade in the intersection be the minimum of the membership grades in the two original fuzzy sets:

$$\mu_{\mathbf{B}\cap\mathbf{L}}(x) = \min\{\mu_{\mathbf{B}}(x), \mu_{\mathbf{L}}(x)\}.$$

Using this intersection operator, we create the membership grades in **B** ∩ **L** element by element:

$$\begin{aligned}
\mu_{\mathbf{B}\cap\mathbf{L}}(a) &= \min\{\mu_{\mathbf{B}}(a), \mu_{\mathbf{L}}(a)\} = \min\{.8, .6\} = .6, \\
\mu_{\mathbf{B}\cap\mathbf{L}}(b) &= \min\{\mu_{\mathbf{B}}(b), \mu_{\mathbf{L}}(b)\} = \min\{.3, 0\} = 0,
\end{aligned}$$

and so forth. The resulting intersection is the fuzzy set of homes that are beautiful and in a good location:

$$\mathbf{B} \cap \mathbf{L} = \{a|.6,\ b|0,\ c|1,\ d|0,\ e|0,\ f|0,\ g|.1\}.$$

The fuzzy set of beautiful and well-constructed homes is obtained in the same way. The resulting intersection is

$$\mathbf{B} \cap \mathbf{C} = \{a|.5,\ b|.3,\ c|.9,\ d|0,\ e|.5,\ f|0,\ g|.1\}.$$

The intersection operator that we used in these examples is the *standard intersection operator*. For fuzzy sets $\mathbf{R}$ and $\mathbf{S}$, this operator, $h(r,s) = \min\{r,s\}$, assigns membership grade to the elements in $\mathbf{R} \cap \mathbf{S}$ as follows:

$$\mu_{\mathbf{R}\cap\mathbf{S}}(x) = \min\{\mu_{\mathbf{R}}(x), \mu_{\mathbf{S}}(x)\}$$

for all $x \in U$. As was the case for the union operation, the standard operator is not always the operator used in applications. Like union operators, all intersection operators must satisfy a basic list of conditions, and there are many functions that satisfy these basic properties. Again, entire classes of nonstandard operators have been defined, but the standard intersection operator is the operator used most frequently. The reasons for this preference are analogous to the reasons why max is the standard union operator. The operators max and min both have certain "nice" properties that other operators do not have. But let's start with the basic properties.

To be called an intersection operator, a function $h : [0,1] \times [0,1] \to [0,1]$ must satisfy the first four of the following properties. Notice the similarities to the properties for union operators; intuitive justification for these conditions are analogous to those for union operators.

1. Function h must satisfy a *boundary condition*: $h(r,1) = r$ for all $r \in [0,1]$.

2. Function h must be *commutative*; that is, $h(r,s) = h(s,r)$ for all $r,s \in [0,1]$. This condition ensures that $\mathbf{R} \cap \mathbf{S} = \mathbf{S} \cap \mathbf{R}$.

3. Function h must be *monotone nondecreasing*; that is, if $r \le r'$ and $s \le s'$, then $h(r,s) \le h(r',s')$. If an element's membership grade increases in either $\mathbf{R}$ or $\mathbf{S}$, then its membership grade in $\mathbf{R} \cap \mathbf{S}$ cannot decrease.

4. Function h must be *associative*; that is, for all $r,s,t \in [0,1]$, $h\big(r, h(s,t)\big) = h\big(h(r,s), t\big)$. If h is associative then the intersection operation is associative: $\mathbf{R} \cap (\mathbf{S} \cap \mathbf{T}) = (\mathbf{R} \cap \mathbf{S}) \cap \mathbf{T}$.

In many applications, it seems reasonable to ask the intersection operator h to satisfy two additional properties.

5. Function h is *continuous*.

6. Function h is *idempotent*; that is, $h(r,r) = r$. This property implies that $\mathbf{R} \cap \mathbf{R} = \mathbf{R}$.

The standard intersection operator earns the right to be called "standard" because of the following theorem.

Theorem 2. *The standard intersection operator is the only function that satisfies all six intersection properties.*

In fact, it can be shown that min is the only function that satisfies intersection properties 1, 2, 3, and 6. This operator is also associative and continuous. The proof of **Theorem 2** is completely analogous to the proof of **Theorem 1**; therefore, we leave the proof as an exercise.

We conclude this section by using the standard intersection operator to find the intersection of our infinite sets **H** and **K**, defined in **Section 4.2**. Recall that **H** is the fuzzy set of numbers close to 100 and **K** is the fuzzy set of numbers close to 150. See **Figure 5** for the graphs of membership functions $\mu_{\mathbf{H}}$ and $\mu_{\mathbf{K}}$. The solid graph in **Figure 7** gives the membership grades for the fuzzy set of *numbers close to both 100 and 150*, $\mathbf{H} \cap \mathbf{K}$.

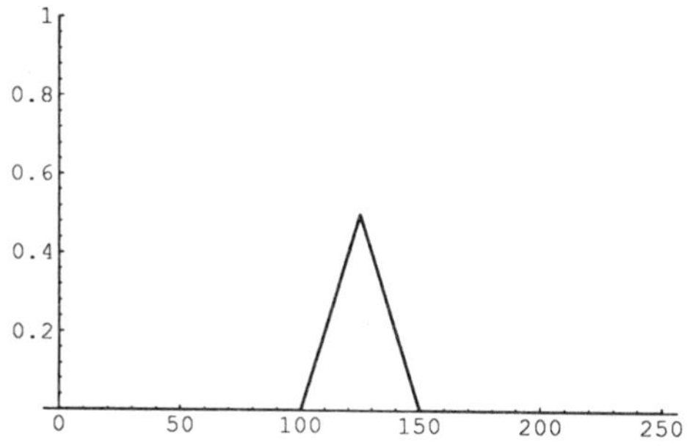

Figure 7. Membership function for $\mathbf{H} \cap \mathbf{K}$, the fuzzy set of numbers close to both 100 and 150.

Algebraically, the membership function for $\mathbf{H} \cap \mathbf{K}$ is given by

$$\mu_{\mathbf{H}\cap\mathbf{K}}(x) = \min\{\mu_{\mathbf{H}}(x), \mu_{\mathbf{K}}(x)\} = \begin{cases} 0, & \text{if } x \in (-\infty, 100); \\ 0.02x - 2, & \text{if } x \in [100, 125]; \\ -0.02x + 3, & \text{if } x \in (125, 150); \\ 0, & \text{if } x \in [150, +\infty). \end{cases}$$

Exercises

13. Recall sets

$$\begin{aligned} \mathbf{T} &= \{\otimes|.3,\ \triangle|0,\ \square|0.8,\ \diamond|.9,\ \nabla|.5\}, \\ \mathbf{V} &= \{x|\tfrac{x}{x+1} : x \geq 1\}, \\ \mathbf{Y} &= \{\otimes|.36,\ \triangle|.09,\ \square|0,\ \diamond|1,\ \nabla|.64\}, \qquad \text{and} \\ \mathbf{Z} &= \{x|\tfrac{1}{x} : x \geq 1\}. \end{aligned}$$

Using the standard intersection operator,

a) find $\mathbf{T} \cap \mathbf{Y}$;

b) find $\mathbf{V} \cap \mathbf{Z}$ and sketch the graph of the membership function.

14. For each $w \in (0, \infty)$, there is a *Yager intersection operator*

$$h_w : [0, 1] \times [0, 1] \to [0, 1]$$

defined by

$$h_w(r, s) = 1 - \min\{1,\ [(1 - r)^w + (1 - s)^w]^{1/w}\}.$$

In terms of fuzzy sets $\mathbf{R}$ and $\mathbf{S}$, for an element $x \in U$, let $r = \mu_{\mathbf{R}}(x)$ and $s = \mu_{\mathbf{S}}(x)$; then

$$\mu_{\mathbf{R} \cap \mathbf{S}}(x) = 1 - \min\{1,\ [(1 - r)^w + (1 - s)^w]^{1/w}\}.$$

a) Using Yager's intersection operators and fuzzy sets $\mathbf{T}$ and $\mathbf{Y}$ from **Exercise 13**, find $\mathbf{T} \cap \mathbf{Y}$ when $w = 0.5$. (Express membership grades to the nearest hundredth.)

b) Repeat part **a** using $w = 2$.

15. In classical set theory, DeMorgan's laws provide two basic relationships among the three classical set operations: For all subsets R and S of a universal set U,

$$\overline{R \cup S} = \overline{R} \cap \overline{S} \qquad \text{and} \qquad \overline{R \cap S} = \overline{R} \cup \overline{S}.$$

a) Prove that when the standard operators are used, DeMorgan's laws also hold in fuzzy set theory. (Hint: For the fuzzy law $\overline{\mathbf{R} \cup \mathbf{S}} = \overline{\mathbf{R}} \cap \overline{\mathbf{S}}$, prove that for all $r, s \in [0, 1]$, we have $1 - \max\{r, s\} = \min\{1 - r, 1 - s\}$.)

b) Show that DeMorgan's laws fail to hold when Yager complement, union, and intersection operators are used with $w = 0.5$. (Hint: Use fuzzy sets $\mathbf{T}$ and $\mathbf{Y}$ from **Exercise 13**.)

16. **a)** Carefully state and prove a lemma for intersection operators that is analogous to the lemma preceding **Theorem 1**.

b) Use your lemma to prove **Theorem 2**.

17. Consider these three fuzzy sets: $\mathbf{H}$, the set of "hot" temperatures; $\mathbf{W}$, the set of "warm" temperatures, and $\mathbf{C}$, the set of "cold" temperatures. The membership functions for element t in these sets might be defined by someone from the Midwest in this way (with temperature t measured in degrees Fahrenheit):

$$\mu_{\mathbf{H}}(t) = \begin{cases} 0, & \text{if } t \in (-\infty, 50); \\ \frac{t}{50} - 1, & \text{if } t \in [50, 100]; \\ 1, & \text{if } t \in (100, +\infty) \end{cases}$$

$$\mu_{\mathbf{W}}(t) = \begin{cases} 0, & \text{if } t \in (-\infty, 50); \\ 1 - \frac{|t-75|}{25}, & \text{if } t \in [50, 100]; \\ 0, & \text{if } t \in (100, +\infty) \end{cases}$$

$$\mu_{\mathbf{C}}(t) = \begin{cases} 1, & \text{if } t \in (-\infty, 25); \\ \frac{75-t}{50}, & \text{if } t \in [25, 75]; \\ 0, & \text{if } t \in (75, +\infty) \end{cases}$$

Using standard operators, answer the following questions.

a) If the temperature $t = 80°$, find the extent to which it is "warm or hot" outside.

b) There is a saying that "temperatures in the desert are either hot or cold, but never warm." Find the membership function for such temperatures, using standard operators and the fuzzy sets **H**, **W**, and **C**, defined above.

18. When we use modifiers, such as "very" or "somewhat" with a characteristic, the characteristic is changed; for example, "very hot" is not the same as "hot." Thus, a new fuzzy set must be established, when a modifier is used. When "very" is used, the membership grades for each element are squared. (Since all membership grades are in $[0, 1]$, squaring them reduces all the membership grades in the interval $(0, 1)$.) For "somewhat," we take the square root of each element's membership grade. (Taking square roots in the designated interval increases the membership grades.)

a) Using **H** from the last problem, find the membership function for the fuzzy set of "very hot" temperatures.

b) On the same axis, graph the membership functions for **H** and **very H**.

c) Using **C** from the last problem, find the membership function for the fuzzy set of "somewhat cold" temperatures.

d) On the same axis, graph the membership functions for **C** and **somewhat C**.

4. Fuzzy Set Theory vs. Traditional Set Theory

In **Section 2**, we noticed a need to redefine equality of sets in the fuzzy context. What about subsets? Because of our knowledge of traditional set theory, we expect certain statements about subsets to be true; for example, we might expect $\mathbf{R} \subseteq \mathbf{R} \cup \mathbf{S}$ to hold for fuzzy sets **R** and **S**. However, every fuzzy set contains all elements of the universal set in question; therefore, the statement $\mathbf{R} \subseteq \mathbf{R} \cup \mathbf{S}$ can no longer merely imply that all elements of **R** are also elements of $\mathbf{R} \cup \mathbf{S}$. In the fuzzy context, the notion of subset must be redefined in terms of membership grades.

Let **R** and **S** be two fuzzy sets lying in the same universal set U. We say **R** is a *subset* of **S** if and only if the membership grade of each element of U in **R** is less than or equal to its membership grade in **S**:

$$\mathbf{R} \subseteq \mathbf{S} \text{ if and only if } \mu_{\mathbf{R}}(x) \leq \mu_{\mathbf{S}}(x) \text{ for all } x \in U.$$

With this definition, the following familiar principle still holds true:

$$\mathbf{R} \subseteq \mathbf{S} \text{ and } \mathbf{S} \subseteq \mathbf{R} \text{ if and only if } \mathbf{R} = \mathbf{S}.$$

Also, with any union operator satisfying the first four axioms, $\mathbf{R} \subseteq \mathbf{R} \cup \mathbf{S}$ holds for all fuzzy sets **R** and **S**. For example, using standard union, for all **R** and **S**, **R** is a subset of $\mathbf{R} \cup \mathbf{S}$ because $\mu_{\mathbf{R}}(x) \leq \max\{\mu_{\mathbf{R}}(x), \mu_{\mathbf{S}}(x)\}$ for all $x \in U$.

Exercises

19. In universal set $U = \{1, 2, 3, 4, 5\}$, let

$$\begin{aligned} \mathbf{R} &= \{1|.5,\ 2|.6,\ 3|.7,\ 4|1,\ 5|.9\}, \\ \mathbf{S} &= \{1|.1,\ 2|.1,\ 3|.7,\ 4|.9,\ 5|.7\}, \\ \mathbf{T} &= \{1|.3,\ 2|.5,\ 3|.7,\ 4|1,\ 5|.8\}. \end{aligned}$$

Determine whether the given statement is true or false. Use the standard operators.

a) $\mathbf{R} \cap \mathbf{S} \subseteq \mathbf{R}$

b) $\mathbf{R} \cap (\mathbf{S} \cup \mathbf{T}) = (\mathbf{R} \cap \mathbf{S}) \cup (\mathbf{R} \cap \mathbf{T})$

c) $\overline{\mathbf{R}} \subseteq \mathbf{R}$

20. Let **R**, **S**, and **T** be fuzzy sets in the same universal set U. Using the standard union and intersection operators, prove the following distributive properties for fuzzy intersection and union.

a) $\mathbf{R} \cap (\mathbf{S} \cup \mathbf{T}) = (\mathbf{R} \cap \mathbf{S}) \cup (\mathbf{R} \cap \mathbf{T})$

b) $\mathbf{R} \cup (\mathbf{S} \cap \mathbf{T}) = (\mathbf{R} \cup \mathbf{S}) \cap (\mathbf{R} \cup \mathbf{T})$

21. Do the relationships in **20a** and **20b** hold for all Yager operators?

22. Let **R** and **S** be fuzzy sets in the same universal set U. Prove that the following statements hold for all union and intersection operators satisfying the first four axioms.

a) $\mathbf{R} \subseteq \mathbf{R} \cup \mathbf{S}$

b) $\mathbf{R} \cap \mathbf{S} \subseteq \mathbf{R}$

There are striking similarities between fuzzy set theory and classical set theory. In fact, many of the definitions and rules in fuzzy set theory have been constructed so that traditional set theory is a special case of fuzzy set theory. The definitions have been formulated so that every crisp set can be interpreted as a fuzzy set. For example, consider a crisp set $T = \{i, j, k\}$ from a universal set $U = \{g, h, i, j, k\}$. Elements i, j, and k are definitely in set T, and elements g and h are definitely not in set T. It follows that T can be expressed as a fuzzy set,

$$\mathbf{T} = \{g|0,\ h|0,\ i|1,\ j|1,\ k|1\}.$$

Both sets T and $\mathbf{T}$ contain the same information. A classical set is just a special type of fuzzy set, a fuzzy set in which only 0 and 1 can appear as membership grades.

The conditions placed on fuzzy set operations guarantee that when all the sets involved are classical sets, the fuzzy operations behave exactly like the classical operations. For example, consider the boundary condition for complement operators. A complement operator $f : [0, 1] \rightarrow [0, 1]$ must satisfy the conditions $f(0) = 1$ and $f(1) = 0$. It follows that, when a crisp set R is interpreted as a fuzzy set $\mathbf{R}$, the fuzzy complement $\overline{\mathbf{R}}$ corresponds to the crisp complement $\overline{R}$.

The boundary conditions placed on union and intersection operators are weaker conditions. For a union operator $g : [0, 1] \times [0, 1] \rightarrow [0, 1]$, we might expect the boundary condition to require $g(0, 0) = 0$ and $g(0, 1) = g(1, 0) = g(1, 1) = 1$. In fact, all four of these statements are true, but we need union conditions 1, 2, and 3 to derive these properties of g. Using only the boundary condition 1, we get only two of the statements: $g(0, 0) = 0$ and $g(1, 0) = 1$. Since g is commutative, we also have $g(0, 1) = g(1, 0) = 1$. Since g is monotone nondecreasing, we have $g(0, 1) \leq g(1, 1)$, and it follows that $g(1, 1) = 1$. Similar remarks hold for the intersection operators.

When all the sets involved are crisp sets, the three fuzzy operators return the same crisp sets we would expect in classical set theory. Fuzzy sets are a generalization of classical sets, and fuzzy set theory is a generalization of classical set theory.

However, when some of the sets involved are truly fuzzy (when membership grades strictly between 0 and 1 appear), we can no longer expect all of the familiar results of classical set theory. For example, in classical set theory, for every set R in a universal set U, we have

$$R \cup \overline{R} = U \qquad \text{and} \qquad R \cap \overline{R} = \emptyset,$$

where $\emptyset$ denotes the empty set.

These two theorems of classical set theory are called the *law of the excluded middle* and the *law of contradiction*, respectively. Despite their importance in classical set theory, these "laws" seldom hold in fuzzy set theory. In fuzzy set theory, we interpret U as a fuzzy set $\mathbf{U} = \{x|1 : x \in U\}$ and the empty set as a fuzzy set $\emptyset = \{x|0 : x \in U\}$. Using the standard operators of fuzzy set theory, if there is even one element of one fuzzy set $\mathbf{R}$ in U that has a membership grade other than 0 or 1, then neither the law of the excluded middle nor the law of contradiction holds. To see why, suppose $\mu_{\mathbf{R}}(x) \in (0, 1)$. Then $\mu_{\overline{\mathbf{R}}}(x) \in (0, 1)$. Therefore, both $\mu_{\mathbf{R}\cup\overline{\mathbf{R}}}(x)$ and $\mu_{\mathbf{R}\cap\overline{\mathbf{R}}}(x)$ lie strictly between 0 and 1. It follows that $\mathbf{R} \cup \overline{\mathbf{R}} \neq \mathbf{U}$ and $\mathbf{R} \cap \overline{\mathbf{R}} \neq \emptyset$.

To illustrate these two inequalities, consider home b from our TinyTown example. Home *b* has membership grades between 0 and 1 in the set of *beautiful* homes as well as in the set of *not beautiful* homes. Thus, home *b* is *beautiful* to some extent, and also *not beautiful* to some extent. Since home *b* is in both **B** and $\overline{\mathbf{B}}$ with membership grades between 0 and 1, it must be in the intersection of **B** and $\overline{\mathbf{B}}$ with a membership grade between 0 and 1. In other words, home *b* is in that "excluded middle" territory between "beautiful" and "not beautiful."

Exercises

23. In universal set $U = \{g, h, i, j, k\}$, let $T = \{i, j, k\}$ and $S = \{h, i\}$. Using the standard operators for the fuzzy complement, union and intersection:

a) Express S as a fuzzy set **S**.

b) Find both $\overline{T}$ and $\overline{\mathbf{T}}$.

c) Find both $S \cup T$ and $\mathbf{S} \cup \mathbf{T}$.

d) Find both $S \cap T$ and $\mathbf{S} \cap \mathbf{T}$.

e) Do any of your answers in parts **b**, **c**, and **d** change if you use Yager operators instead of the standard operators?

24. Let function $h : [0,1] \times [0,1] \to [0,1]$ satisfy the first four conditions for intersection operators. Verify that the following boundary conditions hold.

a) $h(1,1) = 1$

b) $h(0,1) = 0$

c) $h(1,0) = 0$

d) $h(0,0) = 0$

25. Assume that complement and union are defined by the standard operators, and that DeMorgan's law $\overline{\mathbf{R} \cap \mathbf{S}} = \overline{\mathbf{R}} \cup \overline{\mathbf{S}}$ holds. Show that the intersection operator in question is necessarily the standard intersection operator. (Hint: $\mathbf{R} \cap \mathbf{S} = \overline{\overline{\mathbf{R}} \cup \overline{\mathbf{S}}}$.)

5. Building on a Fuzzy Foundation

Virtually every branch of classical mathematics can be (and in most cases has been) "fuzzified." That is, the basic concepts and operators of each branch have been redefined to allow for the inclusion of membership grades. In so doing, the subjectivity of the real world becomes a part of each mathematical model. The results range from arithmetic using fuzzy numbers, to fuzzy calculus, to fuzzy topology. Our discussion up to this point has been directed at the foundation of mathematics, set theory. As each area of mathematics rises from the new fuzzy foundation, it must be looked at in a new way. One of the first blocks to be laid on our fuzzy foundation might be the concept of fuzzy relations.

In this section, we define fuzzy relations and use them to reach a family decision regarding the "best" home to buy in TinyTown. Using this tool, we will be able to look at the seven available homes, as a group, in light of their good and bad qualities. Once we have established the qualities most important to each member of your family, we will be able to find which of the seven homes comes closest to each person's idea of "the perfect home." From there, finding the "best" home, overall, for your family is only a small step away.

5.1 Fuzzy Relations

Let U and V be two universal sets. Elements of the Cartesian product $U \times V$ are ordered pairs (x, y) created by taking x from U and y from V. In traditional set theory, a subset R of $U \times V$ is called a *relation* between U and V; such a relation is often denoted $R(U, V)$ for additional clarity. A crisp relation $R(U, V)$ can be turned into a *fuzzy relation* $\mathbf{R}(U, V)$, abbreviated $\mathbf{R}$, by assigning each element (x, y) of $U \times V$ a membership grade indicating the strength of the relationship between the two coordinates x and y. Instead of being just a set of ordered pairs, a fuzzy relation is a set composed of ordered pairs paired with numbers between 0 and 1. A fuzzy relation contains more information than its corresponding crisp relation.

An illustration would be helpful here, but we save the houses in TinyTown for the exercises. Instead, we propose a new system for rating motion pictures currently showing in TinyTown.

Suppose four people in TinyTown are asked their preferences concerning the content of the movies they most enjoy watching. Let U be the set of four people, and let V be the list of movie content preferences they are asked to consider. Suppose

$$\begin{aligned} U &= \{\text{Ann, Bert, Cathy, Don}\} = \{a, b, c, d\} \quad \text{and} \\ V &= \{\text{reality, humor, sex, violence}\} = \{r, h, s, v\}. \end{aligned}$$

The relation $U \times V$ contains $4 \times 4 = 16$ elements:

$$\begin{aligned} U \times V = \{&(a,r),(a,h),(a,s),(a,v),(b,r),(b,h),(b,s),(b,v), \\ &(c,r),(c,h),(c,s),(c,v),(d,r),(d,h),(d,s),(d,v)\}. \end{aligned}$$

Let a fuzzy relation $\mathbf{P}(U, V)$ assign a membership grade to each ordered pair in $U \times V$ signifying the extent to which that person from U *prefers* that element from V to be emphasized in the movies they see. Since our fuzzy preference relation $\mathbf{P}$ contains all 16 elements of $U \times V$, listing all the ordered pairs with their respective membership grades is cumbersome. A *membership matrix*, $M_{\mathbf{P}}$, provides a concise way of displaying all the membership grades for $\mathbf{P}$. To construct the membership matrix, we assign an order to the elements of U and to the elements of V. The entry in row i and column j of the membership matrix is an element of $[0, 1]$ indicating the degree to which element i of U is related to element j of V. Suppose that in our example we have the following membership matrix for fuzzy relation $\mathbf{P}$:

$$M_{\mathbf{P}} = \begin{array}{c} \\ a \\ b \\ c \\ d \end{array} \begin{array}{c} \begin{array}{cccc} r & h & s & v \end{array} \\ \begin{pmatrix} .9 & 1 & .3 & .2 \\ .6 & .4 & .5 & .6 \\ .6 & .8 & 0 & .3 \\ 0 & .4 & .8 & 1 \end{pmatrix} \end{array}.$$

The row and column labels serve as reminders of the ordering of elements in U and V. The first row shows how Ann (from U) relates to each of the elements of V. Its entries indicate the extent to which Ann prefers movies that emphasize each of the areas. The ordered pair (Ann, reality), or (a, r), is assigned a membership grade of .9 because Ann wants the movies she watches to portray the world the way it is. She also thoroughly enjoys a good comedy, so the ordered pair (Ann, humor), or (a, h), has a membership grade of 1. On the other hand, Don would prefer that his movies not be tied to reality at all. He thoroughly enjoys watching bikini-clad young girls running around a beach as creatures from outer space destroy their entire town. Hence the membership grade for (d, r) is 0 while both (d, s) and (d, v) have high membership grades.

What about the movies currently showing in TinyTown? How are the movies related to the four areas of content in set V? A fuzzy content relation assigns a membership grade to each ordered pair (movie, content) showing the extent to which that movie is associated with that content-type. These membership grades would answer questions that the motion-picture industry's current rating system cannot. How well does this movie portray reality? Does it attempt to be humorous? Does sex (or violence) appear quite often in the film? The rating "R" covers a wide range of movies, as do "G," "PG," and "PG13." Fuzzy relations could greatly enhance the benefits of a motion-picture rating system.

In our example, suppose there are three movies currently showing in TinyTown. Let

$$W = \{\text{the set of movies in town}\} = \{e, f, g\}$$

and, as before, let

$$V = \{\text{reality, humor, sex, violence}\} = \{r, h, s, v\}.$$

Let the fuzzy relation $\mathbf{Q}(V, W)$, the content relation, rate each movie in each of the four content areas. Suppose experts in the motion picture industry give the movies the ratings shown in matrix $M_{\mathbf{Q}}$ below:

$$M_{\mathbf{Q}} = \begin{array}{c} \\ r \\ h \\ s \\ v \end{array} \begin{array}{c} \begin{array}{ccc} e & f & g \end{array} \\ \begin{pmatrix} 0 & .4 & .8 \\ .1 & .9 & 1 \\ .8 & .2 & 0 \\ .9 & .1 & .2 \end{pmatrix} \end{array}.$$

The first column contains the ratings for movie e. Evidently it contains quite a bit of sex and violence (since $\mu_{\mathbf{Q}}(s, e) = .8$ and $\mu_{\mathbf{Q}}(v, e) = .9$), has very little comedy ($\mu_{\mathbf{Q}}(h,e) = .1$), and is not at all concerned about portraying reality ($\mu_{\mathbf{Q}}(r,e) = 0$). The second column has the ratings for movie f, and the third column gives ratings for movie g.

In this section, our examples have focused on fuzzy relations $\mathbf{R}(U,V)$ between small finite sets U and V. When the sets involved are large or infinite, even the matrix description of the fuzzy relation is too cumbersome. Fortunately, it is often possible to find a nice expression for the fuzzy relation's membership function, $\mu_{\mathbf{R}} : U \times V \to [0,1]$. The membership function $\mu_{\mathbf{R}}$ is a function of two variables that completely determines the fuzzy relation $\mathbf{R}$.

Exercises

26. Let U and V be the following sets of states:

$$U = \{\text{Alaska, Colorado, Delaware}\},$$

$$V = \{\text{Colorado, Delaware, Rhode Island, Texas}\}.$$

Let $\mathbf{L}(U,V)$ be the fuzzy relation "x is much larger than y," where $x \in U$, $y \in V$, and "large" refers to area. Use the following information about the areas (in square miles) of the five states to find a suitable choice for the membership matrix $M_{\mathbf{L}}$.

Alaska	Colorado	Delaware	Rhode Island	Texas
570,374	103,730	1,982	1,045	261,914

27. In each of the following parts, a fuzzy relation $\mathbf{R}((0,\infty),(0,\infty))$ is described. Find a suitable choice for $\mu_{\mathbf{R}}(x,y)$.

a) y is much larger than x.

b) y is about x.

5.2 Composition

We can use our two fuzzy relations $\mathbf{P}$ and $\mathbf{Q}$ in **Section 5.1** to draw some conclusions about which movie is most appropriate for which person, but to accomplish this we need a new operation: composition. Given two fuzzy relations, such as $\mathbf{P}$ and $\mathbf{Q}$, the composition operation allows us to combine information contained in the two relations. In our example, the composition of $\mathbf{P}(U,V)$ and $\mathbf{Q}(V,W)$ enables us to combine information about the individual likes and dislikes of the four people in U with information about the ratings for the three movies in W. The result of the composition is a new membership matrix. Entries in this new matrix allow us to draw some fairly accurate conclusions about which movie each person would enjoy most.

The *composition* of two fuzzy relations $\mathbf{P}(U,V)$ and $\mathbf{Q}(V,W)$ is another fuzzy relation, denoted $\mathbf{P} \circ \mathbf{Q}$. Membership levels in this new relation indicate how the elements from the universal set, U, corresponding to rows of $M_{\mathbf{P}}$ relate to the elements from the universal set, W, corresponding to columns of $M_{\mathbf{Q}}$. That is,

$$\mathbf{P}(U,V) \circ \mathbf{Q}(V,W) = (\mathbf{P} \circ \mathbf{Q})(U,W).$$

Notice that the universal set V that contributes elements to the columns of $M_{\mathbf{P}}$ is the same universal set that contributes elements for the rows of $M_{\mathbf{Q}}$.

When calculating the composition of two fuzzy relations, the membership grades in the two matrices are paired up in the same way they would be for multiplication of two matrices, but the product and sum operations are replaced by other operations. Choices for the new operations vary, but in the most common definition of the composition of fuzzy relations, the min operation replaces multiplication and the max operation replaces addition. The *max-min composition* is defined as follows:

> For universal sets U, V, and W and fuzzy relations $\mathbf{P}(U,V)$ and $\mathbf{Q}(V,W)$, the composition $\mathbf{P} \circ \mathbf{Q}$ consists of all ordered pairs in the Cartesian product $U \times W$ with membership grades for elements in the composition given by
>
> $$\mu_{\mathbf{P}\circ\mathbf{Q}}(x,z) = \max_{y\in V}\{\min[\mu_{\mathbf{P}}(x,y), \mu_{\mathbf{Q}}(y,z)]\},$$
>
> for all $x \in U$ and $z \in W$.

That is, to get the membership grade for a given ordered pair (x', z') in the composition, we must look at the membership grades for all ordered pairs $(x', y) \in \mathbf{P}$ and $(y, z') \in \mathbf{Q}$. For each $y \in V$ that is paired with the given $x' \in \mathbf{P}$ and the given $z' \in \mathbf{Q}$, take the minimum of the two membership grades $\mu_{\mathbf{P}}(x', y)$ and $\mu_{\mathbf{Q}}(y, z')$. The largest of all these minimum values, the "max of the mins," is the membership grade for the ordered pair (x', z') in the composition. If V is infinite, the maximum might not exist. In such cases, we use the supremum instead of the maximum.

If we look at composition intuitively, we see why "max-min" is the most commonly used definition for composition. Intuitively, for given elements $x' \in U$ and $z' \in W$, every choice of an element $y \in V$ forms a "bridge" between x' and z': x' is related to z' because x' is related to y and y is related to z'. In the composition, the membership level of (x', z') should be the strength of the strongest bridge connecting x' and z'. Each such bridge is composed of two sections: x' to y and y to z'. These two sections of the bridge have strengths given by $\mu_{\mathbf{P}}(x', y)$ and $\mu_{\mathbf{Q}}(y, z')$, respectively. Since no bridge is stronger than its weakest link or section, $\min\{\mu_{\mathbf{P}}(x', y), \mu_{\mathbf{Q}}(y, z')\}$ is the strength of the bridge determined by our choice of y. The strength of the strongest bridge connecting x' and z' is the maximum value of all such minima.

For example, in **Figure 8** there are three bridges from x' to z'. Each bridge has two sections. For each bridge, the dotted section is the weak section. The strongest of the weak links is from x' to y_2 . Therefore, the bridge that passes through y_2 is the strongest bridge from x' to z' ; this bridge has strength .6.

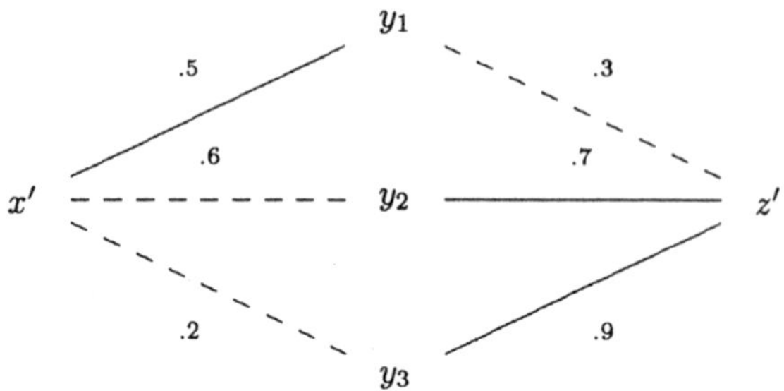

Figure 8. Three bridges from x' to z'.

(If the fuzzy math user does not feel the strength of each bridge should be even as strong as the weakest link or section, then other operations, such as "product," are used to replace "min.")

In our example, we were looking for the movie(s) that each person would most enjoy watching. The composition of fuzzy relations **P** and **Q** gives the desired results. Applying the max-min operation to entries in the membership matrices $M_{\mathbf{P}}$ and $M_{\mathbf{Q}}$, we obtain matrix $M_{\mathbf{P}\circ\mathbf{Q}} = M_{\mathbf{P}} \circ M_{\mathbf{Q}}$ as follows:

$$M_{\mathbf{P}} \circ M_{\mathbf{Q}} = \begin{matrix} \\ a \\ b \\ c \\ d \end{matrix} \begin{matrix} \begin{matrix} r & h & s & v \end{matrix} \\ \begin{pmatrix} .9 & 1 & .3 & .2 \\ .6 & .4 & .5 & .6 \\ .6 & .8 & 0 & .3 \\ 0 & .4 & .8 & 1 \end{pmatrix} \end{matrix} \circ \begin{matrix} \\ r \\ h \\ s \\ v \end{matrix} \begin{matrix} \begin{matrix} e & f & g \end{matrix} \\ \begin{pmatrix} 0 & .4 & .8 \\ .1 & .9 & 1 \\ .8 & .2 & 0 \\ .9 & .1 & .2 \end{pmatrix} \end{matrix} = \begin{matrix} \\ a \\ b \\ c \\ d \end{matrix} \begin{matrix} \begin{matrix} e & f & g \end{matrix} \\ \begin{pmatrix} .3 & .9 & 1 \\ .6 & .4 & .6 \\ .3 & .8 & .8 \\ .9 & .4 & .4 \end{pmatrix} \end{matrix}.$$

To calculate the entries in the composition's membership matrix, we applied the max-min operation element by element. For example,

$$\begin{aligned} \mu_{\mathbf{P}\circ\mathbf{Q}}(a,e) &= \max\{\min\{.9,0\},\min\{1,.1\},\min\{.3,.8\},\min\{.2,.9\}\} \\ &= \max\{0,.1,.3,.2\} = .3, \end{aligned}$$

and

$$\begin{aligned} \mu_{\mathbf{P}\circ\mathbf{Q}}(a,f) &= \max\{\min\{.9,.4\},\min\{1,.9\},\min\{.3,.2\},\min\{.2,.1\}\} \\ &= \max\{.4,.9,.2,.1\} = .9. \end{aligned}$$

In the membership matrix for the composition, entries in row 1 indicate the extent to which Ann will enjoy each of the three movies. By our calculations, she would thoroughly enjoy movie g, but could also really like movie f. Row 2 suggests movies e or g would be the best ones for Bert to see, although with membership of .6 he would not enjoy them as much as Ann would enjoy either movie f or movie g. The results in row 3 indicate Cathy would enjoy movies f or g equally, and row 4 would steer Don to movie e.

Returning to our continuing TinyTown example, we can set up a fuzzy relation **P** that allows us to look simultaneously at all seven homes, along with their individual good and bad qualities. We begin with two universal sets:

$$\begin{aligned} U &= \text{the set of homes for sale} = \{a, b, c, d, e, f, g\}, \qquad \text{and} \\ V &= \text{the set of qualities to look at when buying a home} \\ &= \{\text{beauty, affordability, location, construction}\} \\ &= \{q_1, q_2, q_3, q_4\}. \end{aligned}$$

Let the fuzzy relation **P** indicate the extent to which each home (*a, b, c, d, e, f*, or *g*) can be associated with each quality (q_1 , q_2 , q_3 , or q_4). For any given home, the membership grades of that home in the fuzzy sets **B**, **A**, **L**, and **C** give us the membership grades for the ordered pairs having the given home as the first component. (See **Section 3.3** for these membership grades.) Here are four of the entries in our fuzzy relation **P**:

$$\mathbf{P}(U, V) = \{(a, q_1)|.8,\ (a, q_2)|.4,\ (a, q_3)|.6,\ (a, q_4)|.5,\ \ldots\}.$$

In **Exercise 28**, we ask you to find the membership matrix for **P**.

Before buying a home, you, your spouse, and your two children must individually decide what qualities are most important in your prospective new home. Let fuzzy relation **Q** indicate the extent to which each family member believes a given quality is important. Again, let

$$\begin{aligned} V &= \text{the set of qualities to look at when buying a home} \\ &= \{\text{beauty, affordability, location, construction}\} \\ &= \{q_1, q_2, q_3, q_4\}, \end{aligned}$$

and also let

$$\begin{aligned} W &= \text{the set of family members} \\ &= \{\text{you, your spouse, Mikey, Katie}\} \\ &= \{y, sp, m, k\}. \end{aligned}$$

Suppose that the membership matrix for $\mathbf{Q}(V, W)$ is

$$M_{\mathbf{Q}} = \begin{array}{c} \\ q_1 \\ q_2 \\ q_3 \\ q_4 \end{array} \begin{array}{c} \begin{array}{cccc} y & sp & m & k \end{array} \\ \begin{pmatrix} .6 & .7 & .2 & .9 \\ .9 & .9 & .1 & 0 \\ .7 & .6 & .5 & .5 \\ 1 & .8 & .2 & .1 \end{pmatrix} \end{array}.$$

Exercises

28. Find the membership matrix for $\mathbf{P}(U, V)$ in the TinyTown example.

29. **a)** Find $M_{\mathbf{P}\circ\mathbf{Q}}$ in the TinyTown example. Entries in this matrix indicate the degree to which a particular family member believes that a particular home has the qualities that make a home "perfect."

b) For each family member, which of the seven homes for sale in TinyTown comes closest to being the "perfect" home?

30. Entries in the matrix you found in **Exercise 27a** can be used to reach a family decision about which home comes closest to being the perfect home for the family. Suppose that your family chooses to use a weighted average scheme in which each adult's opinion has twice the weight of a child's opinion. In particular, let $\mathbf{H}$ be the fuzzy set of perfect homes for your family, and for each home $x \in U$, let

$$\begin{aligned}\mu_{\mathbf{H}}(x) &= \tfrac{1}{6}[2\mu_{\mathbf{P}\circ\mathbf{Q}}(x, \text{you}) + 2\mu_{\mathbf{P}\circ\mathbf{Q}}(x, \text{your spouse}) + \\ &\qquad \mu_{\mathbf{P}\circ\mathbf{Q}}(x, \text{Mikey}) + \mu_{\mathbf{P}\circ\mathbf{Q}}(x, \text{Katie})].\end{aligned}$$

The perfect home for your family is the home that has the highest membership grade in $\mathbf{H}$.

a) Find the fuzzy set $\mathbf{H}$.

b) Find the perfect home for your family.

31. Suppose that we are looking for the safest route between three towns (towns a, b, and c) on one side of a mountain range and three cities (cities x, y, and z) on the other side of the mountains. The roads between towns a, b, and c and cities x, y, and z all go through one of two mountain villages (d or e), so there are two routes between any given town and city. Each such route has two parts, with a village acting as a link between a given town and city. **Figure 9** shows the two routes from town a to city x.

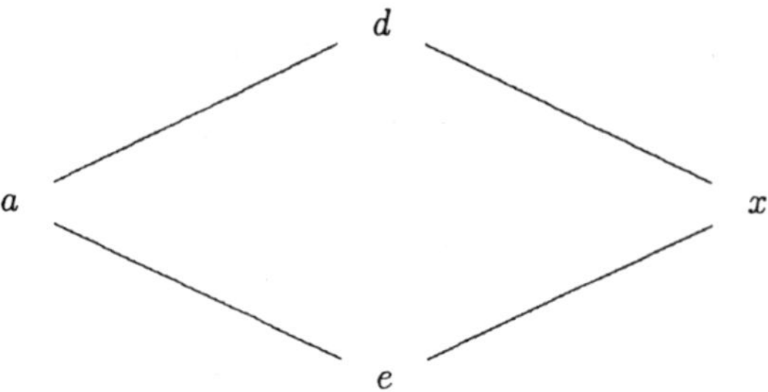

Figure 9. Routes from town a to city x.

Let fuzzy relation $\mathbf{P}$ show the extent to which the routes from towns a, b, and c to villages d and e are safe:

$$M_{\mathbf{P}} = \begin{array}{c} \\ a \\ b \\ c \end{array} \begin{array}{c} \begin{array}{cc} d & e \end{array} \\ \begin{pmatrix} .5 & 0 \\ .4 & .6 \\ .1 & .8 \end{pmatrix} \end{array}.$$

Fuzzy relation **Q** shows the degree of safety for the routes from villages d and e on into cities x, y, and z:

$$M_{\mathbf{Q}} = \begin{array}{c} \\ d \\ e \end{array} \begin{array}{c} \begin{array}{ccc} x & y & z \end{array} \\ \begin{pmatrix} .5 & .9 & .2 \\ .3 & .7 & .6 \end{pmatrix} \end{array}.$$

a) The composition of **P** and **Q** gives the highest degree of safety one can expect, going from a particular town to a given city. Using standard max-min composition, which town and city have the safest route between them?

b) Nonstandard definitions for composition replace min with other functions, like "product." In most applications the results we get when using these replacements are not as appealing as the results we get from "max-min composition." Find the composition of **P** and **Q** using "max-product composition." Compare the degree of safety for the route from town a through village d to city x using the max-min and the max-product compositions. Which composition gives a more reasonable degree of safety for the route and why?

c) In general, how do membership grades in a fuzzy relation found by using max-product composition compare with membership grades in a fuzzy relation found by using max-min composition? Explain why.

6. Conclusion

You might have noticed that the fuzzy set approach to picking the "perfect" home for your family analyzes the problem in much the same way you might have analyzed it in your mind using only common sense. It could be true that our brains instinctively deal with the real world in terms of fuzzy sets, without our knowing it. Solutions for relatively simple problems do not always need to be laid out with the formal structure of fuzzy set theory and all its operations. They often can be handled totally in your brain.

However, fuzzy sets can be very helpful in analyzing more complex problems. Complex problems are seldom "black and white." When "gray" decisions must be made, we want an expert guiding the decision-making process. With fuzzy mathematics, we can take an expert's knowledge about a particular field, organize the knowledge, and record it, so that the expert's knowledge is accessible to others. In fact, the expert's common sense becomes part of a computer program that a novice can use to generate optimal solutions to problems in that field.

In recent years, we, as consumers, have seen a leap in the sophistication of our technological world. Some of the more amazing gizmos and gadgets were engineered using fuzzy mathematics. McNeill and Freiberger [1993] give several interesting examples. We have always known that perfect home videos should not jiggle, but the best "non-fuzzy" solutions to the problem still left us with a jittery image. Fuzzy mathematics has allowed us to attain the desired level of perfection we now enjoy in some video cameras. Using fuzzy mathematics, the focusing mechanism on some cameras instantly compensates for jerky movements. Japanese companies offer their consumers a washing machine engineered with fuzzy mathematics that automatically chooses the best cycle. The user merely puts the clothes in and pushes a start button. The machine makes the subjective decisions the consumer used to make, like the correct water level, how long to run each cycle, etc. These decisions are based on weight of the load, amount of dirt, etc. A subway train in Sendai, Japan, starts and stops with such ease, a passenger must be looking out the window to realize they have departed from one station or arrived at the next. This is done with a 10% savings on fuel, as well. Once again, the engineering is based on fuzzy mathematics, which allows the control system for the train to "decide" the optimal acceleration or deceleration at any given moment.

Fuzzy set theory is a new and exciting area of mathematics. Now that you know the basics, we encourage you to learn more about fuzzy mathematics and its applications. The **References** section contains several texts that are accessible to someone who has studied this Module. In addition to the book by McNeill and Freiberger [1993] mentioned earlier, Kosko [1993] has many examples of applications. Several magazine articles show off the value of fuzzy-mathematics-engineered projects (e.g., Long [1994] and Kosko and Isalea [1993]). For more information about the mathematics of fuzzy set theory, see DuBois and Prade [1980], Klir and Folger [1988], Kosko [1992], and Zimmerman [1985]; in particular, for more about nonstandard operators, see Klir and Folger [1988]. Kosko [1992; 1993] describes an intuitively appealing geometric way of thinking about fuzzy sets. For more information about the mathematics behind the applications of fuzzy set theory, see Klir and Folger [1988], Kosko [1992], Schmucker [1984], and Terano et al. [1992].

7. Solutions to the Exercises

1. a) Solutions will vary, but since the salaries are in descending order, it is reasonable to expect that if $b \in C$ then $a \in C$, and if $c \in C$ then $a, b \in C$, etc. For example, $C = \{a, b\}$ is a possible solution.

b) All four companies are in fuzzy set **F**, but it is reasonable to expect the higher-paying jobs to have higher membership grades in **F**. For example, $\mathbf{F} = \{a|.9,\ b|.7,\ c|.6,\ d|.1\}$ is a possible solution.

2. a) If you were to greedily base your decision solely on the salary offered, you would probably give company *a* probability 1 and the rest probability 0.

b) Numerically, the probabilities and membership grades are probably not the same because they represent different concepts. The probabilities indicate the likelihood of your taking a particular job after considering only salary offers, and the membership grades indicate the extent to which each job pays well. The sum of your membership grades is probably not equal to 1, but the sum of the probabilities must be 1.

3. Solutions will vary. See **Figure 10** for sample solutions.

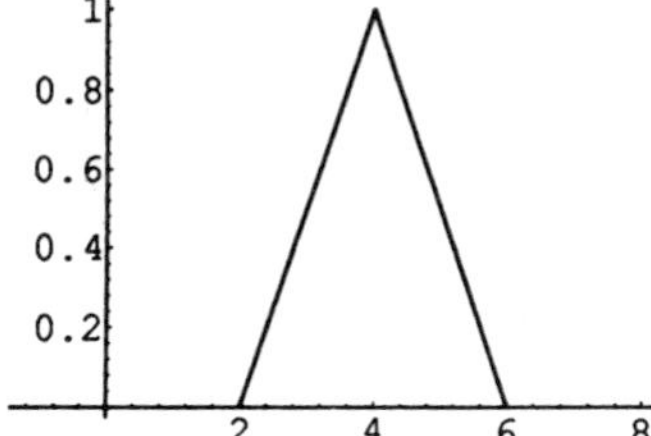

a. Numbers close to 4.

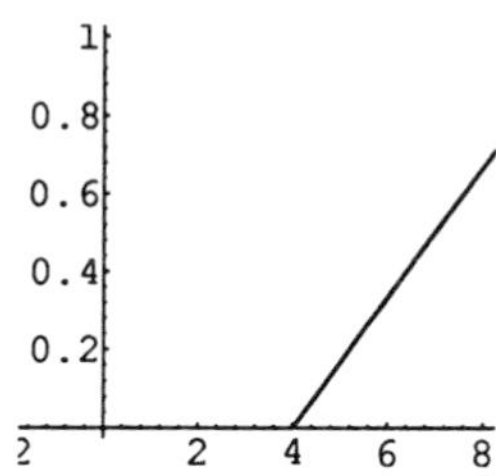

b. Numbers much larger than 4.

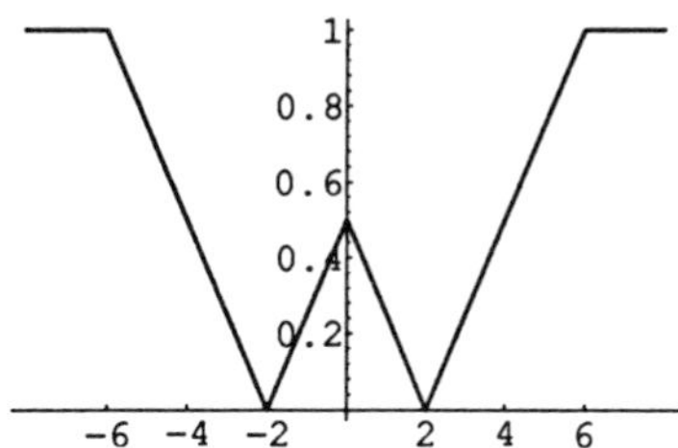

c. Numbers whose squares are not close to 4.

Figure 10. Possible solutions to **Exercise 3**.

4. Solutions will vary. The domain of the membership function is $[0, 100000]$; the codomain is $[0,1]$. It is reasonable to expect a monotone nondecreasing function; that is, higher salaries x should have higher membership grades.

5. a) $\overline{\mathbf{Y}} = \{\otimes|.64,\ \triangle|.91,\ \square|1, \diamondsuit|0,\ \nabla|.36\}$.

b) $\overline{\mathbf{Z}} = \{x|1 - \frac{1}{x} : x \geq 1\}$.

6. a) If $w = .5$, then $\mu_{\overline{\mathbf{R}}}(x) = \left(1 - \sqrt{\mu_{\mathbf{R}}(x)}\right)^2$; therefore,

$$\overline{\mathbf{Y}} = \{\otimes|.16,\ \triangle|.49,\ \square|1,\ \diamondsuit|0,\ \nabla|.04\}, \overline{\mathbf{Z}} = \{x|\left[1 - \left(\tfrac{1}{x}\right)^{.5}\right]^2 : x \geq 1\}.$$

b) If $w = 2$, then $\mu_{\overline{\mathbf{R}}}(x) = \sqrt{1 - \left(\mu_{\mathbf{R}}(x)\right)^2}$; therefore,

$$\overline{\mathbf{Y}} = \{\otimes|.93,\ \triangle|1,\ \square|1,\ \diamondsuit|0,\ \nabla|.77\} \text{ and } \overline{\mathbf{Z}} = \{x|\sqrt{1 - \left(\tfrac{1}{x}\right)^2} : x \geq 1\}.$$

7. a) For all $w > 0$, we have $f_w(0) = (1 - 0^w)^{1/w} = 1$ and $f_w(1) = (1 - 1^w)^{1/w} = 0$.

b) Let $w > 0$ and let $x, y \in [0,1]$ with $x \leq y$. Then $x^w \leq y^w$. It follows that $1 - x^w \geq 1 - y^w$ and $(1 - x^w)^{1/w} \geq (1 - y^w)^{1/w}$. Equivalently, $f_w(x) \geq f_w(y)$.

c) For each $w > 0$, $f_w(r) = (1 - r^w)^{1/w}$ is a composition of continuous functions. Therefore, $f_w(r)$ is also continuous.

d) $f\big(f(r)\big) = (1 - [f(r)]^w)^{1/w} = \left(1 - [(1 - r^w)^{1/w}]^w\right)^{1/w} = \left(1 - (1 - r^w)\right)^{1/w} = (r^w)^{1/w} = r.$

8. Solutions will vary. The functions in **Figure 11** satisfy properties 1 through 4 and are not Yager complements since they are not differentiable.

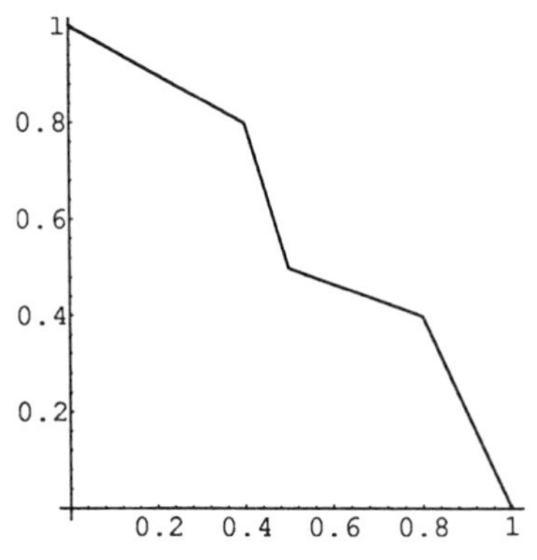

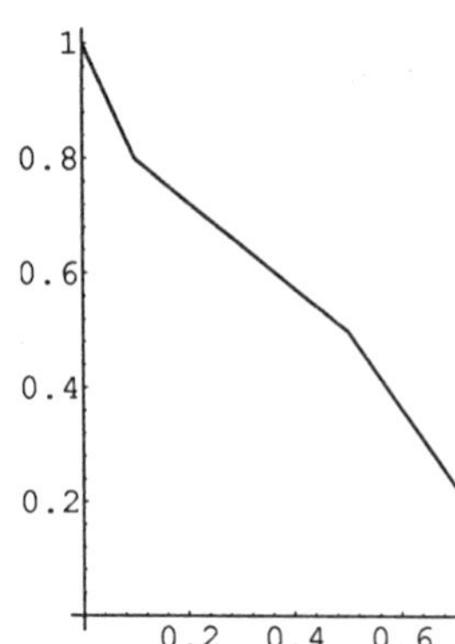

Figure 11. Examples of non-Yager complement operators.

9. a) $\mathbf{T} \cup \mathbf{Y} = \{\otimes|.36,\ \triangle|.09,\ \square|.8,\ \diamondsuit|1,\ \nabla|.64\}$.

b) The elements in $\mathbf{V}$ and $\mathbf{Z}$ are in the interval $[1, \infty)$. The graphs of the membership functions for $\mathbf{V}$ and $\mathbf{Z}$ intersect at the point where $x_0 = (1 + \sqrt{5})/2$. Moreover, $\mu_{\mathbf{Z}}(x) \geq \mu_{\mathbf{V}}(x)$ for all $x \in [0, x_0]$, and $\mu_{\mathbf{V}}(x) \geq \mu_{\mathbf{Z}}(x)$ for all $x \in [x_0, \infty)$. Hence, $\mathbf{V} \cup \mathbf{Z} = \{x|\mu_{\mathbf{V}\cup\mathbf{Z}}(x) : x \geq 1\}$, where

$$\begin{aligned} \mu_{\mathbf{V}\cup\mathbf{Z}}(x) &= \max\{\mu_{\mathbf{Z}}(x), \mu_{\mathbf{V}}(x)\} \\ &= \begin{cases} \mu_{\mathbf{Z}}(x), & \text{if } x \in [1, x_0]; \\ \mu_{\mathbf{V}}(x), & \text{if } x \in (x_0, \infty) \end{cases} \\ &= \begin{cases} \frac{1}{x}, & \text{if } x \in [1, x_0]; \\ \frac{x}{x+1}, & \text{if } x \in (x_0, \infty). \end{cases} \end{aligned}$$

See **Figure 12**.

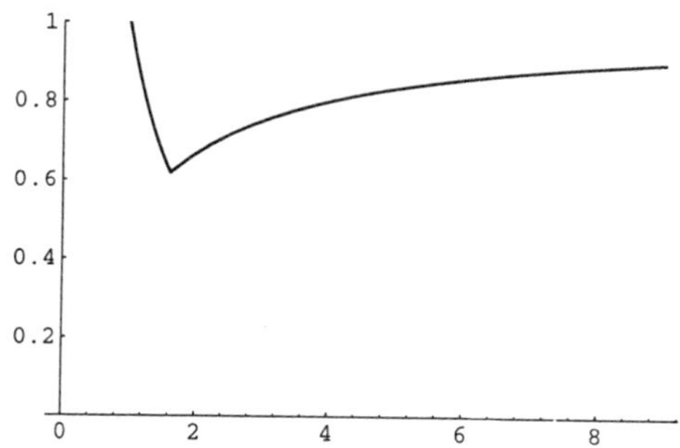

Figure 12. The graph of $\mu_{\mathbf{V}\cup\mathbf{Z}}$ in **Exercise 9b**.

10. a) The function is not idempotent.

b) The function fails to satisfy properties 1 and 4.

c) Properties 1 and 4 do not hold.

d) This function is not idempotent.

11. a) If $w = .5$, then $\mu_{\mathbf{T}\cup\mathbf{Y}}(x) = \min\{1, \left(\sqrt{t} + \sqrt{y}\right)^2\}$, where $t = \mu_{\mathbf{T}}(x)$ and $y = \mu_{\mathbf{Y}}(x)$. Therefore, we have $\mathbf{T} \cup \mathbf{Y} = \{\otimes|1,\ \triangle|.09,\ \square|.8,\ \diamondsuit|1,\ \nabla|1\}$.

b) If $w = 2$, then $\mu_{\mathbf{T}\cup\mathbf{Y}}(x) = \min\{1, \sqrt{t^2 + y^2}\}$. Therefore, $\mathbf{T} \cup \mathbf{Y} = \{\otimes|.47,\ \triangle|.09,\ \square|.8,\ \diamondsuit|1,\ \nabla|.81\}$.

12. a) Let $r \in [0, 1]$ and let $w > 0$. Then $g_w(r, 0) = \min\{1, (r^w + 0^w)^{1/w}\} = \min\{1, r\} = r$.

b) For all $r, s \in [0, 1]$ and all $w > 0$, by commutativity for the real numbers, $r^w + s^w = s^w + r^w$. It follows that $(r^w + s^w)^{1/w} = (s^w + r^w)^{1/w}$; hence, $\min\{1, (r^w + s^w)^{1/w}\} = \min\{1, (s^w + r^w)^{1/w}\}$. Therefore, Yager union operators are commutative.

c) Let $w > 0$ and let $s \in (0,1)$. Then $g_w(s,s) = \min\{1, (s^w + s^w)^{1/w}\} = \min\{1, 2^{1/w}s\} \neq s$.

13. a) $\mathbf{T} \cap \mathbf{Y} = \{\otimes|.3,\ \triangle|0,\ \square|0, \diamondsuit|.9,\ \nabla|.5\}$.

b) The elements in $\mathbf{V} \cap \mathbf{Z}$ are numbers in the interval $[1, \infty)$. In **Exercise 9b**, we found that the graphs of $\mu_{\mathbf{V}}(x) = \frac{x}{x+1}$ and $\mu_{\mathbf{Z}}(x) = \frac{1}{x}$ intersect at the point where $x_0 = (1+\sqrt{5})/2$. Since $\mu_{\mathbf{V}}(x) \leq \mu_{\mathbf{Z}}(x)$ for all $x \in [1, x_0]$ and $\mu_{\mathbf{Z}}(x) \leq \mu_{\mathbf{V}}(x)$ for all $x \in (x_0, \infty)$, we have $\mathbf{V} \cap \mathbf{Z} = \{x | \mu_{\mathbf{V} \cap \mathbf{Z}}(x) : x \geq 1\}$, where

$$\begin{aligned}
\mu_{\mathbf{V}\cap\mathbf{Z}}(x) &= \min\{\mu_{\mathbf{V}}(x), \mu_{\mathbf{Z}}(x)\} \\
&= \begin{cases} \mu_{\mathbf{V}}(x), & \text{if } x \in [1, x_0]; \\ \mu_{\mathbf{Z}}(x), & \text{if } x \in (x_0, \infty) \end{cases} \\
&= \begin{cases} \frac{x}{x+1}, & \text{if } x \in [1, x_0]; \\ \frac{1}{x}, & \text{if } x \in (x_0, \infty). \end{cases}
\end{aligned}$$

See **Figure 13**.

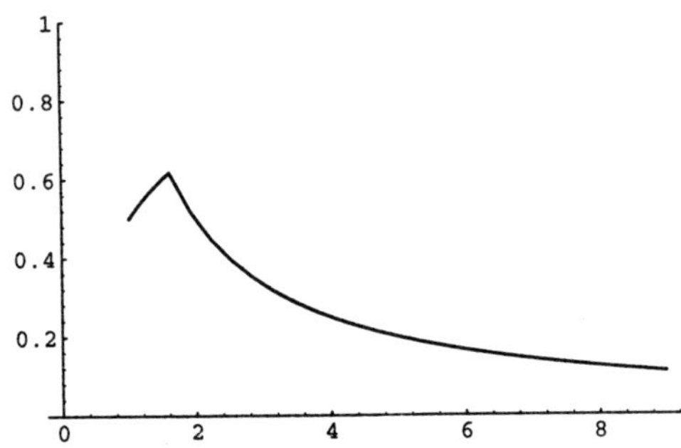

Figure 13. The graph of $\mu_{\mathbf{V}\cap\mathbf{Z}}$ in **Exercise 13b.**

14. a) If $w = .5$, then $\mu_{\mathbf{T}\cap\mathbf{Y}}(x) = 1 - \min\{1, (\sqrt{1-t} + \sqrt{1-y})^2\}$, where $t = \mu_{\mathbf{T}}(x)$ and $y = \mu_{\mathbf{Y}}(x)$. Hence, $\mathbf{T} \cap \mathbf{Y} = \{\otimes|0,\ \triangle|0,\ \square|0,\ \diamondsuit|.9,\ \nabla|0\}$.

b) If $w = 2$, then $\mu_{\mathbf{T}\cap\mathbf{Y}}(x) = 1 - \min\{1, \sqrt{(1-t)^2 + (1-y)^2}\}$. Therefore, $\mathbf{T} \cap \mathbf{Y} = \{\otimes|.05,\ \triangle|0,\ \square|0,\ \diamondsuit|.9,\ \nabla|.38\}$.

15. a) Let $r, s, \in [0,1]$. Assume that $r \leq s$. Then $1 - r \geq 1 - s$. Therefore,

$$\begin{aligned}
1 - \max\{r, s\} &= 1 - s = \min\{1-r, 1-s\}, \qquad \text{and} \\
1 - \min\{r, s\} &= 1 - r = \max\{1-r, 1-s\}.
\end{aligned}$$

b) It suffices to note that $\mu_{\overline{\mathbf{T}\cap\mathbf{Y}}}(\otimes) = 1$ but $\mu_{\overline{\mathbf{T}}\cup\overline{\mathbf{Y}}}(\otimes) = .24$.

16. a) Lemma. *If function $h : [0,1] \times [0,1] \to [0,1]$ satisfies the four intersection properties, then $h(r,s) \leq \min\{r,s\}$ for all $r, s \in [0,1]$.*

Proof: Let $r, s \in [0,1]$. It suffices to show that both $h(r,s) \leq r$ and $h(r,s) \leq s$. Since $s \leq 1$, by the monotone condition, $h(r,s) \leq h(r,1)$. By the boundary condition, $h(r,1) = r$. Combining these two results gives $h(r,s) \leq r$. For the second inequality, first use the commutativity of h, and then use the monotone and boundary conditions on h, to get: $h(r,s) = h(s,r) \leq h(s,1) = s$.

b) Proof: Let $r, s \in [0,1]$. By the lemma, $h(r,s) \leq \min\{r,s\}$. Assume that $r \leq s$. Since h is nondecreasing, $h(r,r) \leq h(r,s)$. Since h is idempotent, $h(r,r) = r$. Consequently, $\min\{r,s\} = r \leq h(r,s)$.

17. a) $\mu_{\mathbf{H}\cup\mathbf{W}}(80) = \max\{\mu_{\mathbf{H}}(80), \mu_{\mathbf{W}}(80)\} = \max\{\frac{3}{5}, \frac{4}{5}\} = \frac{4}{5}$.

b) The set of hot or cold but never warm temperatures is $(\mathbf{H} \cup \mathbf{C}) \cap \overline{\mathbf{W}}$, which has the following membership function:

$$\mu_{(\mathbf{H}\cup\mathbf{C})\cap\overline{\mathbf{W}}}(t) = \begin{cases} 1, & \text{if } t \in (-\infty, 25]; \\ \frac{75-t}{50}, & \text{if } t \in (25, 62.5); \\ \frac{t-50}{50}, & \text{if } t \in [62.5, 66\frac{2}{3}]; \\ \frac{75-t}{25}, & \text{if } t \in (66\frac{2}{3}, 75); \\ \frac{t-75}{25}, & \text{if } t \in [75, 100]; \\ 1, & \text{if } t \in (100, \infty). \end{cases}$$

18. a)

$$\mu_{\text{very hot}}(t) = \begin{cases} 0, & \text{if } t \in (-\infty, 50); \\ \left(\frac{t}{50} - 1\right)^2, & \text{if } t \in [50, 100]; \\ 1, & \text{if } t \in (100, \infty). \end{cases}$$

b) Graphs of the functions $\mu_{\mathbf{H}}(t)$ and $\mu_{\text{very hot}}(t)$. See **Figure 14**.

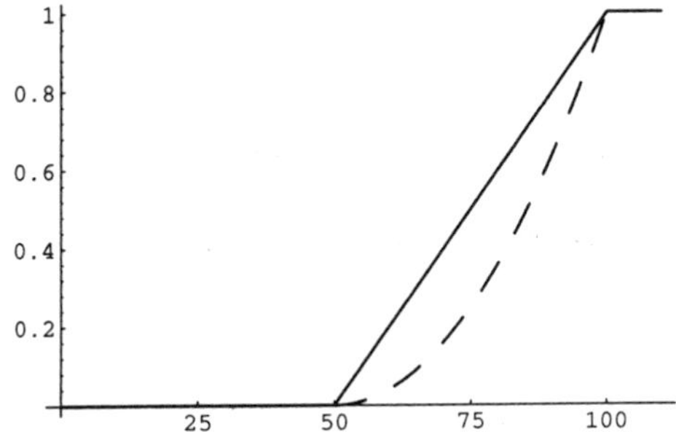

Figure 14. The graphs of $\mu_{\mathbf{H}}$ and $\mu_{\text{very hot}}$ in **Exercise 18b**.

c)

$$\mu_{\text{somewhat cold}}(t) = \begin{cases} 1, & \text{if } t \in (-\infty, 25); \\ \sqrt{\frac{75-t}{50}}, & \text{if } t \in [25, 75]; \\ 0, & \text{if } t \in (75, \infty). \end{cases}$$

d) Graphs of the functions $\mu_{\mathbf{C}}(t)$ and $\mu_{\text{somewhat cold}}(t)$. See **Figure 15.**

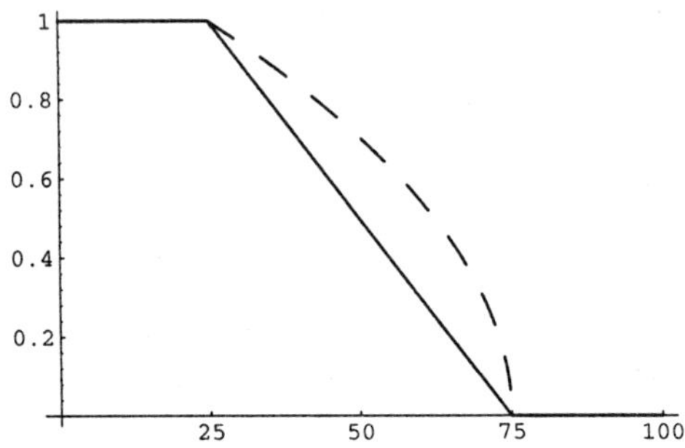

Figure 15. The graphs of $\mu_{\mathbf{C}}$ and $\mu_{\text{somewhat cold}}$ in **Exercise 18d**.

19. All three statements are true.

20. Let r, s, and t be elements of $[0,1]$. For part **a** , it suffices to prove that

$$\min\{r, \max\{s,t\}\} = \max\{\min\{r,s\}, \min\{r,t\}\}.$$

Without loss of generality, we may assume that $s \le t$. It follows that r must satisfy one of the following three inequalities: $r \le s \le t$, $s \le r \le t$, or $s \le t \le r$. Verify that the above equation holds in each of these three cases. The proof of part **b** is similar.

21. No. For example, when $r = s = .3$, $t = .6$, and $w = 2$,

$$\begin{aligned} \mu_{\mathbf{R}\cap(\mathbf{S}\cup\mathbf{T})}(x) &= .2265 \neq .1940 = \mu_{(\mathbf{R}\cap\mathbf{S})\cup(\mathbf{R}\cap\mathbf{T})}(x), \qquad \text{and} \\ \mu_{\mathbf{R}\cup(\mathbf{S}\cap\mathbf{T})}(x) &= .3571 \neq .3368 = \mu_{(\mathbf{R}\cup\mathbf{S})\cap(\mathbf{R}\cup\mathbf{T})}(x). \end{aligned}$$

22. For part **a**, recall that all union operators g satisfy union properties 1 and 3. Therefore, for all r and s in $[0,1]$, we have $r = g(r,0) \le g(r,s)$. It follows that $\mathbf{R} \subseteq \mathbf{R} \cup \mathbf{S}$. The proof in part **b** is similar.

23. a) $\mathbf{S} = \{g|0,\ h|1,\ i|1,\ j|0,\ k|0\}$.

b) $\overline{T} = \{g, h\}$ and $\overline{\mathbf{T}} = \{g|1,\ h|1,\ i|0,\ j|0,\ k|0\}$.

c) $S \cup T = \{h, i, j, k\}$ and $\mathbf{S} \cup \mathbf{T} = \{g|0,\ h|1,\ i|1,\ j|1,\ k|1\}$.

d) $S \cap T = \{i\}$ and $\mathbf{S} \cap \mathbf{T} = \{g|0,\ h|0,\ i|1,\ j|0,\ k|0\}$.

e) No. Yager operators satisfy the conditions that guarantee that fuzzy operators behave like crisp operators when applied to crisp sets.

24. By the boundary condition, $h(1,1) = 1$ and $h(0,1) = 0$. Since h is commutative, $h(1,0) = h(0,1) = 0$. Since h is nondecreasing, we have $h(0,0) \leq h(0,1)$, and it follows that $h(0,0) = 0$.

25. Let f and g represent the standard complement and union operators, respectively. Let h represent an unknown intersection operator. Since $\mathbf{R} \cap \mathbf{S} = \overline{\overline{\mathbf{R}} \cup \overline{\mathbf{S}}}$, for all r and s in $[0,1]$ we have

$$h(r,s) = f\big(g(f(r), f(s))\big) = 1 - \max\{1-r, 1-s\}.$$

Assume that $r \leq s$. Then $h(r,s) = 1-(1-r) = r = \min\{r,s\}$, the standard intersection operator.

26. Solutions will vary.

27. Solutions will vary. Sample solutions follow.

a) $\mu_{\mathbf{R}}(x,y) = \begin{cases} \frac{y-x}{y}, & \text{if } y \geq x; \\ 0, & \text{if } y < x. \end{cases}$

b) $\mu_{\mathbf{R}}(x,y) = 1 - \frac{|x-y|}{x+y}$.

28.

$$M_{\mathbf{P}} = \begin{array}{c} \\ a \\ b \\ c \\ d \\ e \\ f \\ g \end{array} \begin{array}{c} \begin{array}{cccc} q_1 & q_2 & q_3 & q_4 \end{array} \\ \begin{pmatrix} .8 & .4 & .6 & .5 \\ .3 & .7 & 0 & .6 \\ 1 & 0 & 1 & .9 \\ 0 & .9 & .2 & .7 \\ .5 & .4 & 0 & .6 \\ 0 & .8 & 0 & 0 \\ .1 & 1 & .7 & .8 \end{pmatrix} \end{array}.$$

29. a)

$$M_{\mathbf{P} \circ \mathbf{Q}} = \begin{array}{c} \\ a \\ b \\ c \\ d \\ e \\ f \\ g \end{array} \begin{array}{c} \begin{array}{cccc} y & sp & m & k \end{array} \\ \begin{pmatrix} .6 & .7 & .5 & .8 \\ .7 & .7 & .2 & .3 \\ .9 & .8 & .5 & .9 \\ .9 & .9 & .2 & .2 \\ .6 & .6 & .2 & .5 \\ .8 & .8 & .1 & 0 \\ .9 & .9 & .5 & .5 \end{pmatrix} \end{array}.$$

b) Homes c, d, and g come closest to being the perfect home for you. For your spouse, homes d and g are best. Mikey is hard to please, but he would like homes a, c, and g the best. Katie would prefer home c.

30. a) $\mathbf{H} = \{a|.65,\ b|.55,\ c|.80,\ d|.67,\ e|.52,\ f|.55,\ g|.77\}$.

b) Home c comes closest to being the perfect home for your family.

31. a)

$$M_{\mathbf{P}\circ\mathbf{Q}} = \begin{array}{c} \\ a \\ b \\ c \end{array} \begin{array}{c} \begin{array}{ccc} x & y & z \end{array} \\ \begin{pmatrix} .5 & .5 & .2 \\ .4 & .6 & .6 \\ .3 & .7 & .6 \end{pmatrix} \end{array}.$$

Therefore, town c and city y have the safest route between them.

b)

$$M_{\mathbf{P}\circ\mathbf{Q}} = \begin{array}{c} \\ a \\ b \\ c \end{array} \begin{array}{c} \begin{array}{ccc} x & y & z \end{array} \\ \begin{pmatrix} .25 & .45 & .10 \\ .20 & .42 & .36 \\ .24 & .56 & .48 \end{pmatrix} \end{array}.$$

Using the max-product definition, the degree of safety from a to x is .25. However, on both parts of this route, the degree of safety is .5. The .5 degree of safety from a to x given by the max-min definition seems like a more reasonable measure of the degree of safety for the route.

c) The fuzzy relation found by using max-product is always a subset of the fuzzy relation obtained by using max-min, because each product is less than or equal to the corresponding minimum. Thus, the maximums of the products is less than or equal to the maximums of the minimums.

8. Sample Exam

1. In each part, use the fuzzy sets below and the standard definitions for operators to find the requested fuzzy set.

$$\begin{aligned} \mathbf{W} &= \{a|.4,\ b|.1,\ c|0,\ d|.9,\ e|1,\ f|.5\} \\ \mathbf{X} &= \{a|.7,\ b|.2,\ c|.4,\ d|1,\ e|.8,\ f|1\} \\ \mathbf{Y} &= \left\{x|\frac{1}{x+1} : x \geq 1\right\} \\ \mathbf{Z} &= \left\{x|\frac{1}{x} : x \geq 1\right\} \end{aligned}$$

a) $\mathbf{W} \cup \mathbf{X}$

b) $\mathbf{W} \cap \overline{\mathbf{W}}$

c) $\mathbf{Y} \cup \mathbf{Z}$

d) $\mathbf{Z} \cap \overline{\mathbf{Z}}$

2. In classical set theory, given sets A and B, we let $A - B$ denote the set of all elements of A that do not lie in B. It follows that the set difference $A - B$ is equal to the set $A \cap \overline{B}$. For fuzzy sets **A** and **B**, the fuzzy set difference is defined by $\mathbf{A} - \mathbf{B} = \mathbf{A} \cap \overline{\mathbf{B}}$.

   a) Use fuzzy sets **W** and **X** from **Problem 1** and the standard operators to find $\mathbf{W} - \mathbf{X}$.

   b) Give an example of fuzzy sets **A** and **B** such that $\mathbf{A} - \mathbf{B} \subseteq \mathbf{B}$.

3. In **Figure 16**, the dotted graph is the membership function for a fuzzy set **A**, and the solid graph is the membership function for a fuzzy set **B**. Sketch the graph for the membership function of each of the following fuzzy sets.

   a) $\overline{\mathbf{B}}$

   b) $\mathbf{A} \cup \mathbf{B}$

   c) $\mathbf{A} \cap \mathbf{B}$

   d) $\mathbf{B} - \mathbf{A}$

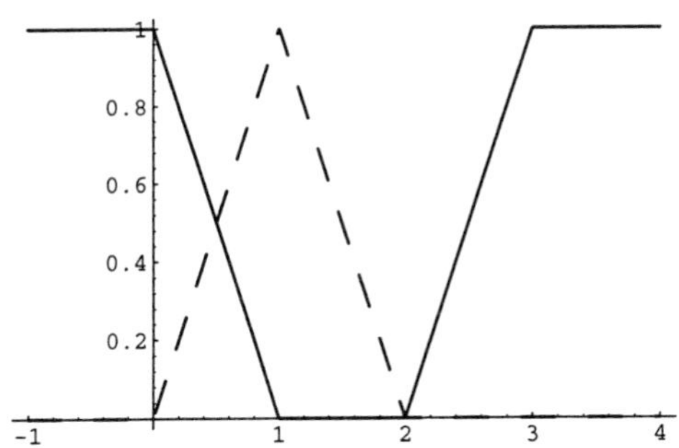

Figure 16. Membership functions for **Problem 3**.

4. Use fuzzy sets **W** and **X** from **Problem 1** and the Yager operators with parameter $w = .5$ to find each of the following fuzzy sets.

   a) $\mathbf{W} \cup \mathbf{X}$

   b) $\overline{\mathbf{W}}$

5. Show that no Yager intersection operator is idempotent.

6. Which of the six intersection properties are not satisfied by the function $h : [0, 1] \times [0, 1] \to [0, 1]$ defined by $h(r, s) = rs$?

7. Let **A** and **B** be fuzzy sets in the same universal set U. Using the standard union operator, prove that $\mathbf{A} \subseteq \mathbf{B}$ if and only if $\mathbf{A} \cup \mathbf{B} = \mathbf{B}$.

8. Consider two universal sets $U = \{3, 4, 5\}$ and $V = \{1, 2, 3, 4\}$. Let $\mathbf{R}(U, V)$ be the fuzzy relation "x is much larger than y," where $x \in U$ and $y \in V$, and let the membership function for **R** be defined as follows:

$$\mu_{\mathbf{R}}(x, y) = \begin{cases} \frac{x-y}{x}, & \text{if } x \geq y; \\ 0, & \text{if } x < y. \end{cases}$$

Find the membership matrix for **R**.

9. Let the membership matrices for two fuzzy relations **P** and **Q** be

$$M_{\mathbf{P}} = \begin{matrix} & x & y & z \\ a & .4 & 0 & 1 \\ b & .2 & 0 & .5 \\ c & .1 & .9 & .1 \\ d & 1 & .4 & .2 \end{matrix}, \qquad M_{\mathbf{Q}} = \begin{matrix} & a & b & c & d \\ x & .1 & 0 & 0 & 0 \\ y & .2 & 0 & .5 & .7 \\ z & 0 & .5 & .6 & 1 \end{matrix}.$$

a) To what extent is c related to z in the fuzzy relation **P**?

b) To what extent is z related to c in fuzzy relation **Q**?

c) To what extent is c related to c in the composition $\mathbf{P} \circ \mathbf{Q}$?

d) Find the membership matrix for $\mathbf{P} \circ \mathbf{Q}$.

9. Solutions to the Sample Exam

1. a) $\mathbf{W} \cup \mathbf{X} = \{a|.7,\ b|.2,\ c|.4,\ d|1,\ e|1,\ f|1\}$.

b) $\overline{\mathbf{W}} = \{a|.6,\ b|.9,\ c|1,\ d|.1,\ e|0,\ f|.5\}$.

c) $\mathbf{Y} \cup \mathbf{Z} = \{x|\frac{1}{x} : x \geq 1\}$.

d) $\mathbf{Z} \cap \overline{\mathbf{Z}} = \{x|\mu_{\mathbf{Z}\cap\overline{\mathbf{Z}}}(x) : x \geq 1\}$ where

$$\mu_{\mathbf{Z}\cap\overline{\mathbf{Z}}}(x) = \begin{cases} 1 - \frac{1}{x}, & \text{if } 1 \leq x \leq 2; \\ \frac{1}{x}, & \text{if } x > 2. \end{cases}$$

2. a) $\mathbf{W} - \mathbf{X} = \{a|.3,\ b|.1,\ c|0,\ d|0,\ e|.2,\ f|0\}$.

b) Solutions will vary; for example, let $\mathbf{A} = \mathbf{B} = \{x|.5,\ y|.5\}$ in universal set $U = \{x, y\}$.

3. See **Figure 17**.

4. a) $\mathbf{W} \cup \mathbf{X} = \{a|1,\ b|.6,\ c|.4,\ d|1,\ e|1,\ f|1\}$.

b) $\overline{\mathbf{W}} = \{a|.1,\ b|.5,\ c|1,\ d|0,\ e|0,\ f|.1\}$.

5. Let $w > 0$. Then $h_w\left(\frac{1}{2}, \frac{1}{2}\right) = 1 - \min\{1,\ 2^{1/w}/2\} \neq \frac{1}{2}$.

6. The function is not idempotent.

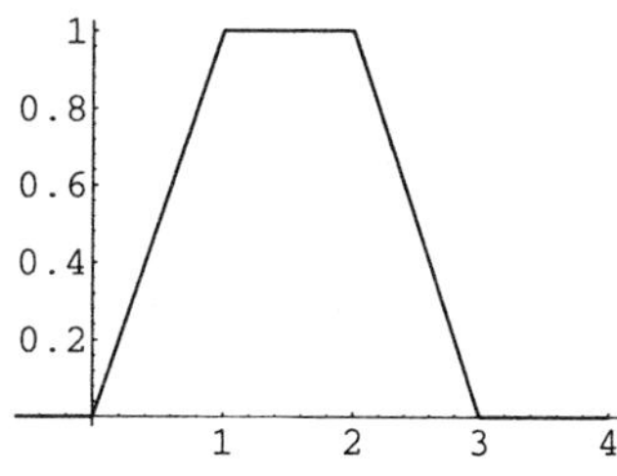

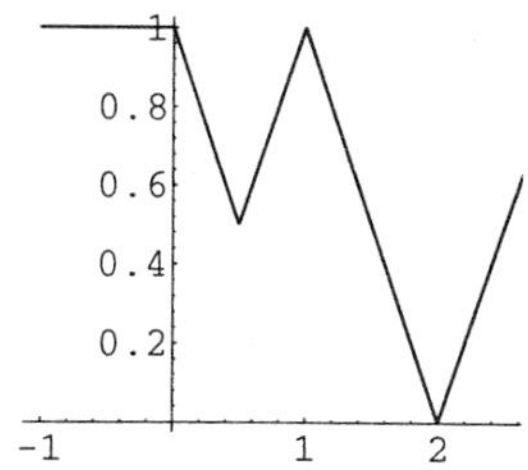

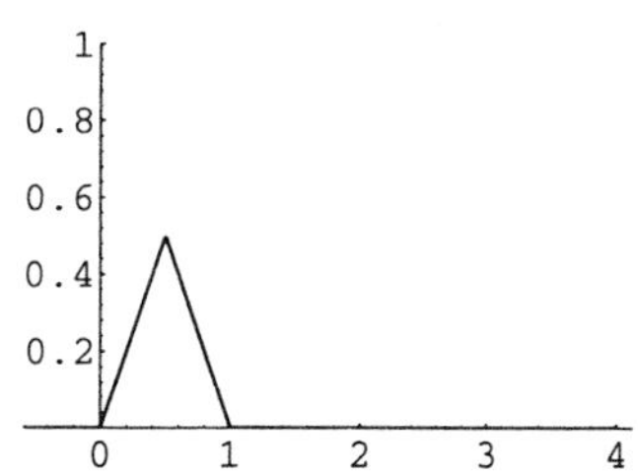

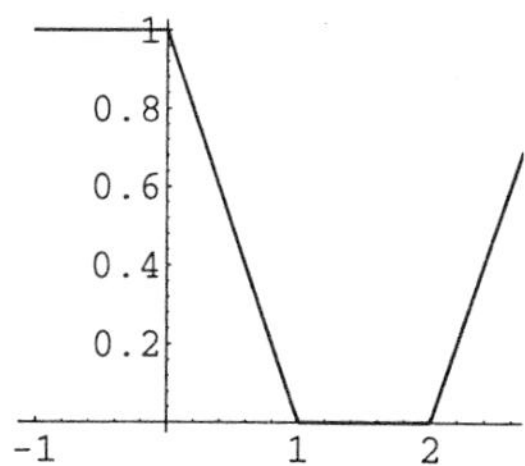

Figure 17. Solutions to **Problem 3.**

7. **Proof:** Notice that for all fuzzy sets $\mathbf{A}$ and $\mathbf{B}$ in some universal set U, we have $\mathbf{A} \subseteq \mathbf{B}$ if and only if for all $x \in U, \mu_{\mathbf{A}}(x) \leq \mu_{\mathbf{B}}(x)$. Moreover, $\mu_{\mathbf{A}}(x) \leq \mu_{\mathbf{B}}(x)$ for all $x \in U$ if and only if for all $x \in U, \max\{\mu_{\mathbf{A}}(x), \mu_{\mathbf{B}}(x)\} = \mu_{\mathbf{B}}(x)$. Finally, the last statement is equivalent to the statement $\mathbf{A} \cup \mathbf{B} = \mathbf{B}$.

8.
$$M_{\mathbf{R}} = \begin{array}{c} \\ 3 \\ 4 \\ 5 \end{array} \begin{array}{c} \begin{array}{cccc} 1 & 2 & 3 & 4 \end{array} \\ \begin{pmatrix} 2/3 & 1/3 & 0 & 0 \\ 3/4 & 2/4 & 1/4 & 0 \\ 4/5 & 3/5 & 2/5 & 1/5 \end{pmatrix} \end{array}.$$

9. **a)** .1

   b) .6

   c) .5

   d)
$$M_{\mathbf{P} \circ \mathbf{Q}} = \begin{array}{c} \\ a \\ b \\ c \\ d \end{array} \begin{array}{c} \begin{array}{cccc} a & b & c & d \end{array} \\ \begin{pmatrix} .1 & .5 & .6 & 1 \\ .1 & .5 & .5 & .5 \\ .2 & .1 & .5 & .7 \\ .2 & .2 & .4 & .4 \end{pmatrix} \end{array}.$$

References

Dubois, Didier, and Henri Prade. 1980. *Fuzzy Sets and Systems: Theory and Applications.* New York: Academic Press.

Klir, George J., and Tina A. Folger. 1988. *Fuzzy Sets, Uncertainty, and Information.* Englewood Cliffs, NJ: Prentice-Hall.

Kosko, Bart. 1992. *Neural Networks and Fuzzy Systems.* Englewood Cliffs, NJ: Prentice-Hall.

________. 1993. *Fuzzy Thinking.* New York: Hyperion.

________, and Satoru Isaka. 1993. Fuzzy logic. *Scientific American* 269 (1) (July 1993) 76–82.

Long, Suzanne. 1994. Fuzzy logic in focus. *Hemispheres* (December 1994) 101–104.

McNeill, Daniel, and Paul Freiberger. 1993. *Fuzzy Logic.* New York: Simon & Schuster.

Schmucker, Kurt J. 1984. *Fuzzy Sets, Natural Language Computations, and Risk Analysis.* Rockville, MD: Computer Science Press.

Terano, Toshiro, Kiyoji Asai, and Michio Sugeno. 1992. *Fuzzy Systems Theory and Its Applications.* San Diego, CA: Academic Press.

Yager, Ron. 1980. On a general class of fuzzy connectives. *Fuzzy Sets and Systems* 4: 235–242.

Zadeh, Lofti. 1965. Fuzzy sets. *Information and Control* 8: 338–353.

Zimmermann, Hans J. 1985. *Fuzzy Set Theory—and Its Applications.* Boston: Kluwer-Nijhoff.

About the Authors

Elizabeth Youngstrom has been a part-time mathematics instructor at Butte Community College for the past seven years. She has a B.A. in mathematics from St. Olaf College in Northfield, Minnesota. She is currently completing her M.S. in mathematics education in the Interdisciplinary Studies Program at California State University, Chico. In the spring of 1994, she was awarded a Graduate Equity Fellowship and chose the writing of this Module as her project.

Margaret Owens has B.A. and M.S. degrees in mathematics from San Jose State University and a Ph.D. in mathematics from the University of Oregon. She is currently a professor of mathematics at California State University, Chico. Her interests include topology, number theory and fuzzy sets. In the Spring of 1994, she served as Elizabeth Youngstrom's mentor and supervised the writing of this Module.